Oscillation Theory for
Neutral Differential Equations
with Delay

Oscillation Theory for Neutral Differential Equations with Delay

D D Bainov

Plovdiv University, Bulgaria

D P Mishev

Technical University, Sofia, Bulgaria

Adam Hilger

Bristol, Philadelphia and New York

British Library Cataloguing-in-Publication Data
Bainov, D.D.
    Oscillation theory for neutral differential
equations with delay.
    I. Title   II. Mishev, D.P.
    515

ISBN 0–7503–0142–2

Library of Congress Cataloging-in-Publication Data
Baĭnov, (Drumi D)
    Oscillation theory for neutral differential equations with delay/
    D.D. Bainov, D.P. Mishev.
        p.   cm.
    Includes bibliographical references and index.
    ISBN 0–7503–0142–2
    1. Differential equations—Delay equations.   2. Oscillations.
    I. Mishev, D.P. (Dimiter P.)   II. Title.
    QA372.B274   1991
    515′.35—dc20   91–15249
                                                           CIP

Published under the Adam Hilger imprint by IOP Publishing Ltd
Techno House, Redcliffe Way, Bristol BS1 6NX, England
335 East 45th Street, New York, NY 10017-3483, USA

Typeset by P & R Typesetters Ltd, Salisbury, Wilts
Printed in Great Britain by Galliard (Printers) Ltd, Great Yarmouth

Contents

Preface

This book is devoted to a rapidly developing branch of the qualitative theory of functional differential equations of neutral type. It will fill a vacuum in the oscillation theory of equations of neutral type and will be a stimulus to its further development.

The book consists of six chapters. After the introduction in Chapter 1, in Chapters 2 and 3 the oscillatory properties of neutral ordinary differential equations of first and second order respectively are investigated. Ordinary differential equations with variable coefficients, with constant coefficients, with distributed delay and nonlinear equations are considered. In Chapter 4 the oscillatory properties of nth order ordinary differential equations of neutral type are investigated and in Chapter 5 systems of neutral type with constant coefficients are considered. Chapter 6 is devoted to the oscillatory properties of partial differential equations of neutral type. Parabolic and hyperbolic equations of neutral type are considered.

The book is addressed to a wide audience of specialists such as mathematicians, physicists, engineers and biologists. It can be used as a textbook at the graduate level and as a reference book for several disciplines.

The authors wish to express their immense thanks to Professor V Kovachev and Professor P Popivanov for helpful comments. We express gratitude to the Ministry of Culture, Science and Education of Bulgaria for its support under Grant 61.

The authors express sincere gratitude to Adam Hilger Publishing House which was so kind as to agree to publish the book proposed by ourselves.

D D Bainov and **D P Mishev**
July 1990

Notation

$\mathbb{N}$	set of all positive integers
$\mathbb{Z}$	set of all integers
$\mathbb{R}$	set of all real numbers
$\mathbb{C}$	set of all complex numbers
$\mathbb{R}^n$	real n-dimensional Euclidean space
$\mathbb{C}^n$	complex n-dimensional Euclidean space
$[a, b]$	closed interval $a \leqslant x \leqslant b$
(a, b)	open interval $a < x < b$
Ω	domain in $\mathbb{R}^n$
$\bar{G}$	closure of the domain G
$C(E; F)$	space of continuous functions $f: E \to F$, $E \subset \mathbb{R}^n$, $F \subset \mathbb{R}$
$C^k(E; F)$	space of functions $f: E \to F$ which are k times continuously differentiable on E

$$f_x = \frac{\partial f}{\partial x}$$

$$x' = \frac{\mathrm{d}x}{\mathrm{d}t}$$

$$\Delta u(x, t) = \sum_{i=1}^{n} u_{x_i x_i}(x, t)$$

1

Introduction

1.1 Preliminary notes

The equations of neutral type to which the present book is devoted play an important role in the theory of functional differential equations. In recent years the theory of this class of equations has become an independent trend and the literature on this subject comprises over 1000 titles. We shall mention the survey on theory of neutral equations by Akhmerov *et al* [1] where a classification is made and a statement of the main problems is given, as well as the book by Kolmanovskii and Nosov [82] in which questions concerning the stability of neutral equations are considered. Many results concerning the theory of neutral functional differential equations were given in the excellent monograph of Hale [66].

The oscillation theory of the solutions of differential equations is one of the traditional trends in the qualitative theory of differential equations. Its essence is to establish conditions for existence of oscillating (nonoscillating) solutions, to study the laws of distribution of the zeros, the maxima and minima of the solution, to obtain estimates of the distance between the neighbouring zeros and the number of zeros in a given interval, to describe the relationship between the oscillatory and other basic properties of the solutions of various classes of differential equations, etc.

The development of oscillation theory for ordinary differential equations began in the 1840s when the classical work of Sturm [147] appeared, in which theorems of oscillation and comparison of the solutions of second order linear homogeneous ordinary differential equations were proved. The first oscillation results for differential equations with a translated argument were obtained by Fite [35] in 1921. It is worth mentioning that already Fite paid attention to the great differences between the oscillatory properties of the solutions of differential equations with a translated argument and of the corresponding equations without a translation of the argument. In recent years the numbers of investigations devoted to the oscillation theory of functional differential equations has considerably increased. A survey of a large number of the works published by 1977 can be found in the monograph of Shevelo [141] where about 400 titles are cited. The recently published monograph of Ladde *et al* [109] citing more than 300 titles is devoted to the systematic study of the oscillatory properties of the solutions of ordinary differential equations with a translated argument. We shall also mention the monographs of Reid [134], Myshkis [123], Norkin [128], Koplatadze and Chanturia [84] and Domshlak [21].

The neutral equations find numerous applications in natural sciences and technology but, as a rule, they enjoy specific properties which make their study difficult both in aspects of ideas and techniques. These difficulties explain the relatively small number of works devoted to the investigation of the oscillatory properties of the solutions of neutral equations. In 1977 Norkin [127] published a paper concerning the oscillation theory of neutral functional differential equations. The heavy restrictions imposed in it, however, practically eliminate the influence of the neutral member. The first work in which a criterion for oscillation of the solutions of neutral equations was proved, essentially different from the classical criteria, was published in 1980 by Zahariev and Bainov [160]. Some general approaches to the investigation of the oscillatory and asymptotic properties of the solutions of neutral equations were given by Bainov *et al* [9], [13], Myshkis *et al* [125], Györi [64] and Ntouyas and Sficas [129]. Sufficient conditions for oscillation and asymptotic behaviour of the solutions of neutral linear differential equations with variable coefficients were obtained by Grammatikopoulos, Ladas and Sficas [51], Grammatikopoulos, Ladas and Meimaridou [50], [49], [48], Ladas and Sficas [101], [99], Grammatikopoulos, Grove and Ladas [47], Zhang [162], Grove *et al* [56], [55], Jaroš and Kusano [78], Erbe and Zhang [30] and Ruan [135], [136], [138]. Theorems on the distribution of the zeros of solutions of neutral equations were obtained by Domshlak [22], Domshlak and Sheikhzamanova [23], [24], Sheikhzamanova [140], Aliev [2] and Domoshnitskii [18–20]. Necessary and sufficient conditions for oscillation of the solutions of differential equations and systems of neutral type with constant coefficients were obtained by Sficas and Stavroulakis [139], Grove and Ladas [57], Kulenović *et al* [94], Grammatikopoulos, Sficas and Stavroulakis [52], Farrel [32], Zhanyuan [163], Grammatikopoulos and Stavroulakis [53], [54], Ladas and Partheniadis [107], Ladas *et al* [108] and Arino and Györi [5]. Sufficient conditions for oscillation and asymptotic behaviour of the solutions of neutral differential equations with constant coefficients were obtained by Ladas and Sficas [101], [100], Grammatikopoulos, Grove and Ladas [45], [46], [47], Grove *et al* [58], Stavroulakis [146], Zhang [162] and Partheniadis [130].

Sufficient conditions for oscillation of the solutions of systems of differential equations of neutral type were obtained by Shevelo *et al* [143] and Györi and Ladas [65]. A necessary and sufficient condition for oscillation of the solutions of first order neutral equations with distributed delay was obtained by Bainov *et al* [8]. Sufficient conditions for oscillation of the solutions of neutral equations with distributed delay were obtained by Bainov *et al* [11], [12] and Györi [64]. Sufficient conditions for oscillation and asymptotic behaviour of the solutions of nonlinear differential equations of neutral type were obtained by Jaroš and Kusano [75], [76], [77], Ivanov [72], Ivanov and Kusano [73], [74], Ladas [97], Graef *et al* [43], [44], Arino and Bourad [4], Erbe and Zhang [31], Grace and Lalli [40], [41], [42], Zahariev and Bainov [158], [159] and Drakhlin [25]. Oscillatory properties of the solutions of a class of neutral equations with 'maxima' were investigated by Bainov and Zahariev [14]. Sufficient conditions for oscillation of the solutions of neutral integro-differential equations were obtained by Bainov *et al* [10] and Ruan [137]. Oscillatory properties of the solutions of the neutral logistic equation with delay were investigated by Györi [62], [63] and Gopalsamy and Zhang [39].

The development of oscillation theory for partial differential equations began in the 1960s with the appearance of the works of Protter [133], Kreith [86], [90] and Headley and Swanson [67]. A survey of the results published by 1973 can be found in the monograph of Kreith [87].

In the past few years the fundamental theory of partial differential equations with a translated argument has been developed intensively. The qualitative theory of these important classes of partial differential equations is, however, still in an initial stage of development. Thus, for instance, a small number of works have now been published since 1983 on the theory of oscillations for this class of equations. We shall mention the works of Georgiou and Kreith [36], Kreith and Ladas [92], Yoshida [153], [156], Bykov and Kultaev [17], Tramov [149] and Mishev [117]. The first paper in which a criterion for oscillation of the solutions of partial differential equations of neutral type was given was published in 1984 by Mishev and Bainov [122]. Sufficient conditions for oscillation of the solutions of neutral hyperbolic equations were obtained by Mishev and Bainov [119], [121] and Mishev [114]. Theorems on the distribution of the zeros of the solutions of nonlinear hyperbolic equations of neutral type were proved by Yoshida [154]. Criteria for oscillation of the solutions of neutral parabolic equations were obtained by Mishev and Bainov [120] and, for nonlinear parabolic equations of neutral type, by Mishev [115]. Neutral hyperbolic and parabolic equations with 'maxima' were investigated by Mishev [113], [116]. Necessary and sufficient conditions for oscillation of the solutions of some classes of hyperbolic and parabolic equations of neutral type were obtained by Mishev and Bainov [118] and Mishev [112].

1.2   Auxiliary assertions

Consider the ordinary differential equation

$$x' = f(t, x) \tag{1.2.1}$$

together with the initial condition

$$x(t_0) = x_0. \tag{1.2.2}$$

It is well known that under certain assumptions about f the initial value problem (1.2.1), (1.2.2) has a unique solution and it is equivalent to the integral equation

$$x(t) = x(t_0) + \int_{t_0}^{t} f(s, x(s))\, ds, \quad t \geq t_0. \tag{1.2.3}$$

Next, consider a differential equation of the form

$$\frac{dx(t)}{dt} = f(t, x(t), x(t - \tau)), \quad \tau > 0, \quad t \geq t_0 \tag{1.2.4}$$

where the right-hand side depends not only on the instantaneous position $x(t)$, but also on $x(t - \tau)$, the position at τ units back, that is to say, the equation has past memory. Such an equation is called an ordinary differential equation with a delay or retarded argument. Whenever necessary, we shall consider the integral equation

$$x(t) = x(t_0) + \int_{t_0}^{t} f(s, x(s), x(s - \tau))\, ds \tag{1.2.5}$$

which is equivalent to (1.2.4). In order to define a solution of (1.2.4), we need to have a known function $\phi(t)$ on $[t_0 - \tau, t_0]$, instead of just the initial condition $x(t_0) = x_0$.

The basic initial value problem for an ordinary differential equation with delay argument is posed as follows: on the interval $[t_0, T]$, $T \leq +\infty$, we seek a continuous function x that satisfies (1.2.4) and an initial condition

$$x(t) = \phi(t), \quad t \in E_{t_0} \tag{1.2.6}$$

where t_0 is an initial point, $E_{t_0} = [t_0 - \tau, t_0]$ is the initial set; the known function $\phi(t)$ on E_{t_0} is called the initial function. Usually, it is assumed that $x(t_0 + 0) = \phi(t_0)$. We always mean a one-sided derivative when we speak of the derivative at an end-point of an interval.

Under general assumptions, the existence and uniqueness of solutions to the initial value problem (1.2.4), (1.2.6) can be established. The solution is sometimes denoted by $x(t, \phi)$. In the case of a variable delay $\tau = \tau(t) > 0$ in equation (1.2.4), it is also required to find a solution of this equation for $t > t_0$ such that on the initial set $E_{t_0} = t_0 \cup \{t - \tau(t): t - \tau(t) < t_0, \ t \geqq t_0\}$, $x(t)$ coincides with the given initial function $\phi(t)$. If it is required to determine the solution on the interval $[t_0, T]$, then the initial set $E_{t_0 T}$ is $\{t_0\} \cup \{t - \tau(t): t - \tau(t) < t_0, t_0 \leqq t \leqq T\}$.

Example 1.2.1 For the equation

$$x'(t) = f(t, x(t), x(t - \cos^2 t)) \tag{1.2.7}$$

$t_0 = 0$, $E_0 = [-1, 0]$, and the initial function $\phi(t)$ must be given on the interval $[-1, 0]$.

The initial set E_{t_0} depends on the initial point t_0. This statement can be justified by the following example.

Example 1.2.2 For the equation

$$x'(t) = ax(t/2) \tag{1.2.8}$$

$E_0 = \{0\}$ and $E_1 = [1/2, 1]$.

Now consider the differential equation of nth order with l deviating arguments, of the form

$$x^{(m_0)}(t) = f(t, x(t), \ldots, x^{(m_0 - 1)}(t),$$
$$x(t - \tau_1(t)), \ldots, x^{(m_1)}(t - \tau_1(t)), \ldots,$$
$$x(t - \tau_l(t)), \ldots, x^{(m_l)}(t - \tau_l(t))) \tag{1.2.9}$$

where the deviations $\tau_i(t) > 0$, and $\mu = \max_{1 \leq i \leq l} m_i$.

In order to formulate the initial value problem for (1.2.9), we shall need the following notation. Let t_0 be the given initial point. Each deviation $\tau_i(t)$ defines the initial set $E_{t_0}^{(i)}$ given by

$$E_{t_0}^{(i)} = \{t_0\} \cup \{t - \tau_i(t): t - \tau_i(t) < t_0, \ t \geqq t_0\}.$$

We denote $E_{t_0} = \cup_{i=1}^{l} E_{t_0}^{(i)}$, and on E_{t_0} continuous functions $\phi_k(t)$, $k = 0$, $1, \ldots, \mu$, must be given, with $\mu = \max_{1 \leq i \leq l} m_i$. In applications, it is most natural to consider the case where on E_{t_0},

$$\phi_k(t) = \phi_0^{(k)}(t), \quad k = 0, 1, 2, \ldots, \mu$$

but it is not generally necessary.

The nth order differential equation should be given initial values $x_0^{(k)}$, $k = 0, 1, 2, \ldots, n-1$. Now let $x_0^{(k)} = \phi_k(t_0)$, $k = 0, 1, \ldots, \mu$. If $\mu < n-1$, then, in addition, the numbers $x_0^{(\mu+1)}, \ldots, x_0^{(n-1)}$ are given. If the point t_0 is an isolated point of E_{t_0}, then $x_0^{(0)}, \ldots, x_0^{(n-1)}$ are also given.

For equation (1.2.9), the basic initial value problem consists of the determination of an $(n-1)$-times continuously differentiable function x that satisfies equation (1.2.9) for $t > t_0$ and the conditions

$$x^{(k)}(t_0 + 0) = x_0^{(k)}, \quad k = 0, 1, \ldots, n-1$$

$$x^{(k)}(t - \tau_i(t)) = \phi_k(t - \tau_i(t)), \quad \text{if } t - \tau_i(t) < t_0$$

$$k = 0, 1, \ldots, \mu; \quad i = 1, 2, \ldots, l.$$

At a point $t_0 + (k-1)\tau$ the derivative $x^{(k)}(t)$, generally speaking, is discontinuous, but the derivatives of lower order are continuous.

Example 1.2.3   Consider the equation

$$x''(t) = f(t, x(t), x'(t), x(t - \cos^2 t), x(t/2)) \qquad (1.2.10)$$

For $t_0 = 0$, we have $n = 2$, $l = 2$, $\mu = 0$, the initial set $E_0^{(1)}$ is an interval $-1 \leq t \leq 0$, $E_0^{(2)} = \{0\}$, and $E_0 = [-1, 0]$, on which the initial function $\phi_0(t)$: $x_0^{(0)} = \phi_0(0)$ is given and $x_0^{(1)}$ is any given number.

We shall give the following definition of ordinary functional differential equations of neutral type with delay. In the case when $m_0 = \mu$, equation (1.2.9) is said to be an equation of neutral type with delay. We shall not dwell on the definitions of the other classes of ordinary functional differential equations since they are not an object of study of this book.

Example 1.2.4   The equations

$$x'(t) + a(t)x'(t - \tau) + b(t)x(t) = 0, \quad \tau > 0 \qquad (1.2.11)$$

and

$$x''(t) + a(t)x''(t - \tau) + F(t, x(t), x'(t)) = 0, \quad \tau > 0 \qquad (1.2.12)$$

are of neutral type, while the equation

$$x'(t) + c(t)x(t - \tau) = 0, \quad \tau > 0 \qquad (1.2.13)$$

is not of neutral type.

Before we define oscillation of solutions, let us consider some simple examples.

Example 1.2.5   The equation

$$y'' + y = 0$$

has periodic solutions $x(t) = \cos t, \quad y(t) = \sin t.$

Example 1.2.6   Consider the equation

$$y''(t) - (1/t)y'(t) + 4t^2 y(t) = 0 \qquad (1.2.14)$$

whose solution is $y(t) = \sin t^2$. This solution is not periodic but it enjoys an oscillatory property.

Example 1.2.7   Consider the equation

$$y''(t) + (1/2)y(t) - (1/2)y(t - \pi) = 0, \quad t \geqslant 0 \qquad (1.2.15)$$

whose solution $y(t) = 1 - \sin t$ has an infinite sequence of multiple zeros. This solution also enjoys an oscillatory property.

Example 1.2.8   Consider the equation

$$y''(t) - y(-t) = 0 \qquad (1.2.16)$$

which has an oscillating solution $y_1(t) = \sin t$ and a nonoscillating solution $y_2(t) = e^t + e^{-t}$.

Let us now restrict our discussion to those solutions $y(t)$ of the equation

$$y''(t) + a(t)y(t - \tau(t)) = 0 \qquad (1.2.17)$$

which exist on some ray $[T_y, \infty)$ and satisfy $\sup\{|y(t)|: t \geq T\} > 0$ for every $T \geq T_y$. In other words, $|y(t)| \not\equiv 0$ on any infinite interval $[T, \infty)$. Such a solution sometimes is said to be a regular solution.

We usually assume that $a(t) \geq 0$ or $a(t) \leq 0$ in (1.2.17), and in doing so we mean to imply that $a(t) \not\equiv 0$ on any infinite interval $[T, \infty)$.

There are various definitions for the oscillation of solutions of ordinary differential equations (with or without deviating arguments). In this section, we give two definitions of oscillation, which are used in the rest of the book; these are the ones most frequently used in the literature.

As we see from the above examples, the definition of oscillation of regular solutions can have two different forms.

Definition 1.2.1 A nontrivial solution $y(t)$ (implying a regular solution) is said to oscillate if and only if it has arbitrarily large zeros for $t \geq t_0$, that is, there exists a sequence of zeros $\{t_n\}$ $(y(t_n) = 0)$ of $y(t)$ such that $\lim_{n \to \infty} t_n = +\infty$. Otherwise, $y(t)$ is said to be nonoscillating.

For nonoscillating solutions there exists t_1 such that $y(t) \neq 0$, for all $t \geq t_1$.

Definition 1.2.2 A nontrivial solution $y(t)$ is said to oscillate if it changes sign on (T, ∞), where T is any number.

When $\tau(t) \equiv 0$ and $a(t)$ is continuous in (1.2.17), the two definitions given

above are equivalent. This is because of the fact that the uniqueness of the solution makes multiple zeros impossible. However, as Example 1.2.7 suggests, a differential equation with translated arguments can have solutions with multiple zeros. These two definitions are different, especially for higher order ordinary differential equations which may have solutions with multiple zeros.

Definition 1.2.1 is more general than Definition 1.2.2. The solution $y(t) = 1 - \sin t$ of equation (1.2.15) oscillates according to Definition 1.2.1 and is nonoscillating according to Definition 1.2.2.

In example 1.2.7, the possibility of multiple zeros of nontrivial solution is a consequence of the retardation, since if $\tau(t) \equiv 0$, the corresponding equation has no solutions with multiple zeros.

For the system of first order equations with translated arguments

$$\begin{vmatrix} x'(t) = f_1(t, x(t), x(\tau_1(t)), y(t), y(\tau_2(t))) \\ y'(t) = f_2(t, x(t), x(\tau_1(t)), y(t), y(\tau_2(t))) \end{vmatrix} \tag{1.2.18}$$

the solution $(x(t), y(t))$ is said to be strongly (weakly) oscillating if each (at least one) of its components oscillates.

Consider the hyperbolic differential equation of neutral type

$$\frac{\partial^2}{\partial t^2} [u(x, t) + \lambda(t)u(x, t - \tau)] - [\Delta u(x, t) + \mu(t)\Delta u(x, t - \sigma)]$$

$$+ c(x, t, u(x, t)) = f(x, t), \quad (x, t) \in \Omega \times (0, \infty) \equiv G \qquad (1.2.19)$$

where $\tau, \sigma = \text{const.} > 0$, $\Delta u(x, t) = \sum_{i=1}^{n} u_{x_i x_i}(x, t)$ and Ω is a bounded domain in $\mathbb{R}^n$ with a piecewise smooth boundary. Consider boundary conditions of the form

$$\frac{\partial u}{\partial n} + \gamma(x, t)u = 0, \quad (x, t) \in \partial\Omega \times [0, \infty), \qquad (1.2.20)$$

$$u = 0, \quad (x, t) \in \partial\Omega \times [0, \infty). \qquad (1.2.21)$$

Definition 1.2.3 The solution $u(x, t) \in C^2(G) \cap C^1(\bar{G})$ of problems (1.2.19), (1.2.20) ((1.2.19), (1.2.21)) is said to oscillate in the domain G if for any number $\mu \geq 0$ there exists a point $(x_0, t_0) \in \Omega \times [\mu, \infty)$ such that the equality $u(x_0, t_0) = 0$ holds.

2

First order neutral ordinary differential equations

2.1   First order linear differential equations

In this present section the oscillatory properties and asymptotic behaviour of the nonoscillating solutions of first order ordinary differential equations of the form

$$\frac{d}{dt}[y(t) + P(t)y(t - \tau)] + Q(t)y(t - \sigma) = 0, \quad t \geq t_0 \qquad (2.1.1)$$

are investigated, where τ, $\sigma = \text{const.} \geq 0$.

Assume the following conditions are fulfilled:

H2.1.1 $P(t), Q(t) \in C([t_0, \infty); \mathbb{R})$,

H2.1.2 $Q(t) \geq 0$ for $t \geq t_0$, $Q(t) \not\equiv 0$,

H2.1.3 $Q(t) \geq q > 0$ for $t \geq t_0$, where $q = \text{const.}$

Let $\phi(t) \in C([t_0 - \rho, t_0]; \mathbb{R})$, where $\rho = \max\{\tau, \sigma\}$.

Definition 2.1.1   The function $y(t) \in C([t_0 - \rho, \infty); \mathbb{R})$ is said to be a solution of equation (2.1.1) with initial function $\phi(t)$ if $y(t) = \phi(t)$ for $t \in [t_0 - \rho, t_0]$, the function $y(t) + P(t)y(t - \tau)$ is continuously differentiable for $t \geq t_0$ and $y(t)$ satisfies equation (2.1.1).

Definition 2.1.2   The solution $y(t)$ of equation (2.1.1) defined in the interval $[T_y, \infty)$, where $T_y \geq t_0$, is said to be a regular solution if for any number $T \geq T_y$ the following condition holds

$$\sup\{|y(t)| : t \geq T\} > 0.$$

Definition 2.1.3 The regular solution $y(t)$ of equation (2.1.1) is said to oscillate if the function $y(t)$ has a sequence of zeros tending to $+\infty$. Otherwise the solution is said to be nonoscillating.

Definition 2.1.4 The function $\varphi(t): [t_\varphi, \infty) \to \mathbb{R}$ is said to eventually enjoy the property S if there exists an interval $[\bar{t}_\varphi, \infty) \subset [t_\varphi, \infty)$ in which $\varphi(t)$ enjoys the property S.

The subsequent theorems are devoted to the asymptotic behaviour of the nonoscillating solutions of equation (2.1.1). In the proof of the theorems we shall use the following lemma:

Lemma 2.1.1

Let $y(t)$ be an eventually positive solution of equation (2.1.1) and

$$z(t) = y(t) + P(t)y(t - \tau).$$

Then the following assertions are valid:
(a) If conditions H2.1.1, H2.1.2 hold, then $z(t)$ is a monotone decreasing function.
(b) If conditions H2.1.1, H2.1.2 hold as well as the condition

$$-1 < p_1 \leqq P(t), \tag{2.1.2}$$

where $p_1 = $ const., then

$$z(t) > 0. \tag{2.1.3}$$

(c) If conditions H2.1.1, H2.1.2 hold in addition to

$$-1 < p_1 \leqq P(t) \leqq 0 \tag{2.1.4}$$

and

$$\int_{t_0}^{\infty} Q(s) \, ds = \infty \tag{2.1.5}$$

then

$$\lim_{t \to \infty} z(t) = 0. \tag{2.1.6}$$

(*d*) *If conditions H2.1.1, H2.1.3 hold in addition to the condition*

$$P(t) \leqq p_2 < -1 \qquad (2.1.7)$$

where $p_2 = $ const., *then*

$$z(t) < 0. \qquad (2.1.8)$$

(*e*) *If conditions H2.1.1, H2.1.3 hold in addition to the condition*

$$p_1 \leqq P(t) \leqq p_2 \qquad (2.1.9)$$

then

$$\lim_{t \to \infty} z(t) = -\infty \quad \text{or} \quad \lim_{t \to \infty} z(t) = 0.$$

(*f*) *If conditions H2.1.1, H2.1.3 hold in addition to the condition*

$$p_1 \leqq P(t) \leqq p_2 < -1 \qquad (2.1.10)$$

then

$$\lim_{t \to \infty} z(t) = -\infty. \qquad (2.1.11)$$

(*g*) *If conditions H2.1.1, H2.1.3 hold in addition to the condition*

$$-1 \leqq P(t) \leqq p_2 \qquad (2.1.12)$$

then

$$\lim_{t \to \infty} z(t) = 0. \qquad (2.1.13)$$

Proof

(*a*) From equation (2.1.1) and condition H2.1.2 it follows that

$$z'(t) = -Q(t)y(t - \sigma) \leqq 0. \qquad (2.1.14)$$

(*b*) Suppose that inequality (2.1.3) does not hold. Since $z(t)$ is a monotone function, then

$$z(t) \equiv 0 \quad \text{for } t \geqq t_1 \geqq t_0 \qquad (2.1.15)$$

or

$$z(t) < 0 \quad \text{for } t \geqq t_1 \geqq t_0. \tag{2.1.16}$$

If (2.1.15) holds, then from equation (2.1.1) it follows that $Q(t)y(t - \tau) \equiv 0$. Since $y(t) > 0$, then $Q(t) \equiv 0$, which contradicts condition H2.1.2.

If (2.1.16) holds, then from (2.1.2) it follows that

$$y(t) < -P(t)y(t - \tau) \leqq -p_1 \cdot y(t - \tau).$$

From the above inequality we obtain that $p_1 \in (-1, 0)$ and $-1 < P(t) < 0$. Then

$$y(t + n\tau) \leqq (-p_1)^n y(t) \to 0, \quad n \to \infty.$$

Hence $\lim_{t \to \infty} y(t) = 0$ and since the function $P(t)$ is bounded, then $\lim_{t \to \infty} z(t) = 0$. From the last equality we obtain that $z(t) > 0$, which contradicts inequality (2.1.16). Thus assertion (b) is proved.

(c) From assertions (a) and (b) it follows that $z(t)$ is a positive and decreasing function. Hence $\lim_{t \to \infty} z(t) = l \geqq 0$. Suppose that $l > 0$. Since $z(t) \leqq y(t)$, then from (2.1.14) we obtain

$$z'(t) + (l/2)Q(t) \leqq 0. \tag{2.1.17}$$

Integrate both sides of (2.1.17) over the interval $[t_1, t]$ and obtain

$$\frac{l}{2} \int_{t_1}^{t} Q(s) \, ds \leqq z(t_1) - z(t).$$

Hence $\int_{t_0}^{\infty} Q(s) \, ds < \infty$, which contradicts condition (2.1.5).

(d) Suppose that (2.1.8) is not valid. Then

$$z(t) = y(t) + P(t)y(t - \tau) \geqq 0. \tag{2.1.18}$$

Hence

$$y(t) \geqq -P(t)y(t - \tau) \geqq -p_2 y(t - \tau)$$

and

$$y(t + n) \geqq (-p_2)^n y(t) \to \infty, \quad n \to \infty.$$

Then

$$\lim_{t \to \infty} y(t) = \infty. \tag{2.1.19}$$

Since

$$z'(t) = -Q(t)y(t - \tau) \leq -qy(t - \tau) \qquad (2.1.20)$$

then from (2.1.19) we obtain that $\lim_{t \to \infty} z'(t) = -\infty$, from which it follows that $\lim_{t \to \infty} z(t) = -\infty$ too. But the last equality contradicts inequality (2.1.18). Thus assertion (d) is proved.

  (e) From inequality (2.1.20) it follows that the function $z(t)$ is monotone decreasing. Hence $\lim_{t \to \infty} z(t) = -\infty$ or $\lim_{t \to \infty} z(t) = l \in \mathbb{R}$. Suppose that the second equality holds. Integrate both sides of (2.1.20) over the interval $[t_1, t]$ and obtain

$$0 \leq \int_{t_1}^{t} qy(s - \sigma)\, ds \leq z(t_1) - l.$$

Hence $y(t) \in L_1[t_1, \infty)$. In view of (2.1.9) we obtain that $z(t) \in L_1[t_1, \infty)$. This is possible only if $l = 0$, that is,

$$\lim_{t \to \infty} z(t) = 0.$$

  (f) Suppose that (2.1.11) is not satisfied. Then from assertion (e) it follows that

$$\lim_{t \to \infty} z(t) = 0$$

from which we obtain that

$$z(t) > 0. \qquad (2.1.21)$$

Hence

$$y(t) > -P(t)y(t - \tau) \geq -p_2 y(t - \tau)$$

and

$$y(t + n\tau) \geq (-p_2)^n y(t) \to \infty, \quad n \to \infty.$$

From the last inequality we obtain that $\lim_{t \to \infty} y(t) = \infty$ and from inequality (2.1.20) we obtain that $\lim_{t \to \infty} z'(t) = -\infty$. Hence $\lim_{t \to \infty} z(t) = -\infty$. But the last equality contradicts inequality (2.1.21).

(*g*) Suppose that (2.1.13) does not hold. Then from assertion (*e*) it follows that

$$\lim_{t \to \infty} z(t) = -\infty \qquad (2.1.22)$$

hence

$$z(t) < 0.$$

Then $y(t) < -P(t)y(t - \tau) \leq y(t - \tau)$ which implies that $y(t)$ is a bounded function. This fact contradicts (2.1.22). Thus assertion (*g*) is proved.

Theorem 2.1.1

Assume that condition H2.1.3 and condition (2.1.10) hold. Then each non-oscillating solution $y(t)$ of equation (2.1.1) tends to $-\infty$ or $+\infty$ as $t \to \infty$.

Proof

Let $y(t)$ be an eventually positive solution of equation (2.1.1). (We shall note that if $y(t)$ is an eventually negative solution of (2.1.1), then $-y(t)$ is an eventually positive solution of the same equation.) Introduce the notation $z(t) = y(t) + P(t)y(t - \tau)$. Then from Lemma 2.1.1 (*f*) we obtain that $\lim_{t \to \infty} z(t) = -\infty$. Hence

$$p_1 y(t - \tau) \leq P(t)y(t - \tau) < z(t) \to -\infty, \quad t \to \infty$$

from which we obtain that

$$\lim_{t \to \infty} y(t) = \infty.$$

Theorem 2.1.1 is proved.
   The following example illustrates Theorem 2.1.1.

Example 2.1.1   Consider the equation

$$\frac{d}{dt}[y(t) + (-2 + e^{-2t})y(t - \tfrac{1}{2}\ln 2)]$$

$$+ [2(\sqrt{2} - 1) + \sqrt{2}\,e^{-2t}]y(t - \ln 2) = 0, \quad t \geq 1. \qquad (2.1.23)$$

A straightforward verification shows that all conditions of Theorem 2.1.1 are met. Hence each nonoscillating solution $y(t)$ of equation (2.1.23) tends to $+\infty$ or $-\infty$ as $t \to \infty$. For instance, $y(t) = e^t$ and $y(t) = -e^t$ are such solutions.

The following example illustrates the fact that if the coefficient $P(t)$ is not bounded below, then Theorem 2.1.1 is not valid.

Example 2.1.2   Consider the equation

$$\frac{d}{dt}[y(t) - e^t\, y(t-1)] + e^{-2}\, y(t-2) = 0, \quad t \geqq 0. \qquad (2.1.24)$$

A straightforward verification yields that condition H2.1.3 holds but condition (2.1.10) does not hold since $P(t) = -e^t$ is not a bounded below function. We shall note that the function $y(t) = e^{-t}$ is a solution of equation (2.1.24) for which $\lim_{t \to \infty} y(t) = 0$.

Theorem 2.1.2

Let condition H2.1.3 hold in addition to one of the conditions

$$0 \leqq P(t) \leqq p_2 \qquad (2.1.25)$$

or

$$-1 < p_1 \leqq P(t) \leqq 0, \qquad (2.1.26)$$

where p_1, p_2 are constants. Then each nonoscillating solution $y(t)$ of equation (2.1.1) tends to zero as $t \to \infty$.

Proof

Let $y(t)$ be an eventually positive solution of equation (2.1.1). Introduce the notation $z(t) = y(t) + P(t)y(t-\tau)$. It is immediately verified that the conditions of Lemma 2.1.1 (g) hold. Hence

$$\lim_{t \to \infty} z(t) = 0. \qquad (2.1.27)$$

Suppose that condition (2.1.25) holds. Then using (2.1.27) and the inequality $z(t) \geqq y(t) > 0$ we obtain that $\lim_{t \to \infty} y(t) = 0$. Suppose that condition (2.1.26) holds. First we shall prove that $y(t)$ is a bounded solution. From

(2.1.27) it follows that $z(t) > 0$. Since $z(t)$ is a decreasing function, then $z(t) \leqq B$, where $B = \text{const.}$, hence

$$y(t) \leqq -P(t)y(t - \tau) + B \leqq -p_1 y(t - \tau) + B. \qquad (2.1.28)$$

Suppose that $y(t)$ is not a bounded function. In this case there exists a sequence $\{t_n\}_{n=1}^{\infty}$ such that

$$\lim_{n \to \infty} t_n = \infty, \quad \lim_{n \to \infty} y(t_n) = \infty \text{ and } y(t_n) = \max_{t_0 \leqq s \leqq t_n} y(s).$$

Using (2.1.28), for n large enough we obtain

$$y(t_n) \leqq -p_1 y(t_n - \tau) + B \leqq -p_1 y(t_n) + B$$

that is, $(1 + p_1)y(t_n) \leqq B$ which contradicts the assumption that $y(t)$ is not a bounded function. Hence $\lim_{t \to \infty} \sup y(t) = s$, where $s \in \mathbb{R}$. Let $\{t_n\}_{n=1}^{\infty}$ be a sequence of numbers such that $\lim_{n \to \infty} y(t_n) = s$. Then for n large enough we obtain

$$z(t_n) = y(t_n) + P(t_n)y(t_n - \tau) \geqq y(t_n) + p_1 y(t_n - \tau)$$

that is

$$y(t_n - \tau) \geqq -(1/p_1)[-z(t_n) + y(t_n)].$$

From the last inequality for $n \to \infty$ we obtain that

$$s \geqq \lim_{n \to \infty} \sup y(t_n - \tau) \geqq (s/-p_1).$$

Since $-p_1 \in (0, 1)$, then $s = 0$. Thus Theorem 2.1.2 is proved.

Example 2.1.3   Consider the equation

$$\frac{\mathrm{d}}{\mathrm{d}t}[y(t) + (\tfrac{1}{2} + e^{-t})y(t - \tfrac{1}{2}\ln 2)] + \left(\frac{2 + \sqrt{2}}{4} + \sqrt{2}\,e^{-t}\right)$$

$$\times y(t - \ln 2) = 0, \quad t \geqq 0. \qquad (2.1.29)$$

A straightforward verification yields that all conditions of Theorem 2.1.2 are met. Hence each nonoscillating solution of equation (2.1.29) tends to zero as $t \to \infty$. For instance, $y(t) = e^{-t}$ is such a solution.

Theorem 2.1.3

Let condition H2.1.2 hold in addition to one of the conditions

$$P(t) \geqq 0 \tag{2.1.30}$$

or

$$-1 < p_1 \leqq P(t) \leqq 0. \tag{2.1.31}$$

Then the following assertions are valid:
(a) Each nonoscillating solution of equation (2.1.1) is bounded.
(b) If the condition

$$\int_{t_0}^{\infty} Q(s)\, ds = \infty \tag{2.1.32}$$

holds and $y(t)$ is a nonoscillating solution of equation (2.1.1), then

$$\liminf_{t \to \infty} y(t) = 0. \tag{2.1.33}$$

(c) If conditions (2.1.31) and (2.1.32) hold, then each nonoscillating solution of equation (2.1.1) tends to zero as $t \to \infty$.

Proof

Let $y(t)$ be an eventually positive solution of equation (2.1.1). Introduce the notation $z(t) = y(t) + P(t)y(t - \tau)$. From Lemma 2.1.1 (a) and (b) it follows that

$$\lim_{t \to \infty} z(t) = l \geqq 0. \tag{2.1.34}$$

(a) Suppose that condition (2.1.30) holds. Then

$$z(t) = y(t) + P(t)y(t - \tau) \geqq y(t).$$

From (2.1.34) it follows that the solution $y(t)$ is bounded. If condition (2.1.31) holds, then by arguments analogous to those in the proof of Theorem 2.1.2 we obtain that $y(t)$ is a bounded function.

(b) Since $z'(t) = -Q(t)y(t - \sigma)$, from (2.1.34) we obtain that $Q(t)y(t - \sigma) \in L_1[t_0, \infty)$. In view of condition (2.1.32) this is possible only if equality (2.1.33) holds.

(c) From Lemma 2.1.1 (a), (b) and (c) it follows that $z(t)$ is a decreasing function, $z(t) = y(t) + P(t)y(t - \tau) \geq 0$ and $\lim_{t \to \infty} z(t) = 0$. Hence $z(t) \leq B$, where $B = \text{const}$. By arguments analogous to those in the proof of Theorem 2.1.1 we obtain that $\lim_{t \to \infty} y(t) = 0$.

Example 2.1.4 Consider the equation

$$\frac{d}{dt}[y(t) + e^t y(t - 1)] + \tfrac{1}{2}y(t - \ln 2) = 0, \quad t \geq 0. \quad (2.1.35)$$

A strightforward verification yields that all conditions of Theorem 2.1.3 hold. Hence each nonoscillating solution $y(t)$ of equation (2.1.35) is bounded and enjoys the property (2.1.33). For instance, $y(t) = e^{-t}$ is such a solution.

Example 2.1.5 Consider the equation

$$\frac{d}{dt}[y(t) + (-\tfrac{1}{2} + e^{-t})y(t - 1)] + \frac{(2 + 2c - e)e^t + 4e}{2(c\,e^{2t} + e^{t+2})}$$

$$\times\, y(t - 2) = 0, \quad t \geq 1 \quad (2.1.36)$$

where $c \geq \tfrac{1}{2}e - 1$. A straightforward verification yields that the conditions of Theorem 2.1.3 are satisfied, except for condition (2.1.32). We shall note that $y(t) = c + e^{-t}$ is a nonoscillating solution of equation (2.1.36) for which

$$\lim_{t \to \infty} y(t) = c \neq 0.$$

In the following theorems sufficient conditions for oscillation of the solutions of equation (2.1.1) are given. In the proof of Theorems 2.1.4 and 2.1.5 the following lemma is used:

Lemma 2.1.2

Assume that the following condition holds

$$\liminf_{t \to \infty} \int_{t-\mu}^{t} p(s)\,ds > 1/e$$

where μ = const. > 0 and $p(t) \in C([t_0, \infty); (0, \infty))$. Then
  (i) The differential inequality

$$x'(t) - p(t)x(t + \mu) \geq 0, \quad t \geq t_0$$

has no eventually positive solutions.
  (ii) The differential inequality

$$x'(t) - p(t)x(t + \mu) \leq 0, \quad t \geq t_0$$

has no eventually negative solutions.
  (iii) The differential inequality

$$x'(t) + p(t)x(t - \mu) \leq 0, \quad t \geq t_0$$

has no eventually positive solutions.
  (iv) The differential inequality

$$x'(t) + p(t)x(t - \mu) \geq 0, \quad t \geq t_0$$

has no eventually negative solutions.

Proof

(i) Suppose that the assertion is not true. Let $x(t)$ be a solution of the differential inequality such that $x(t) > 0$ for $t \geq t_1 \geq t_0$. Then $x'(t) \geq p(t)x(t + \mu) > 0$, from which it follows that $x(t) < x(t + \mu)$ for $t \geq t_1$. Introduce the notation

$$w(t) = \frac{x(t + \mu)}{x(t)}.$$

Hence $\lambda = \lim_{t \to \infty} \inf w(t) \geq 1$. Divide both sides of the differential inequality in (i) by $x(t)$ and obtain

$$\frac{x'(t)}{x(t)} - p(t)\frac{x(t + \mu)}{x(t)} \geq 0.$$

$$\ln w(t) \geq \int_t^{t+\mu} p(s)w(s)\, ds. \tag{2.1.37}$$

Case I. Let $\lambda \in \mathbb{R}$. From (2.1.37) we obtain

$$\ln \lambda \geq \lambda \lim_{t \to \infty} \inf \int_t^{t+\mu} p(s)\, ds.$$

Using the fact that

$$\max_{x \geq 1}(\ln x - ax) = -\ln a - 1$$

from the last inequality it follows that

$$\max_{\lambda \geq 1}\left[\ln \lambda - \lambda \lim_{t \to \infty} \inf \int_t^{t+\mu} p(s)\, ds\right]$$

$$= -\ln\left(\lim_{t \to \infty} \inf \int_t^{t+\mu} p(s)\, ds\right) - 1 \geq 0.$$

Hence

$$\ln\left(\lim_{t \to \infty} \inf \int_t^{t+\mu} p(s)\, ds\right) \leq -1,$$

$$\lim_{t \to \infty} \inf \int_t^{t+\mu} p(s)\, ds \leq 1/e.$$

The last inequality contradicts the condition of Lemma 2.1.2.

Case II. Let $\lambda = \infty$, that is

$$\lim_{t \to \infty} \frac{x(t + \mu)}{x(t)} = +\infty. \tag{2.1.38}$$

From the condition of Lemma 2.1.2 it follows that

$$\int_t^{t+\mu} p(s)\, ds \geq B > 1/e \quad \text{for } t \geq t_2 \geq t_1 \tag{2.1.39}$$

where $B = \text{const.}$ From (2.1.39) it follows that for any $t \geq t_2 + \mu$ we can find a number t^* such that

$$\int_t^{t^*} p(s)\,ds \geq B/2, \qquad \int_{t^*-\mu}^t p(s)\,ds \geq B/2 \qquad (2.1.40)$$

for $t^* - \mu < t < t^*$. Integrating both sides of the differential inequality in (*i*) over the interval $[t, t^*]$ and using (2.1.40) we obtain

$$x(t^*) - x(t) \geq \int_t^{t^*} p(s)x(s + \mu)\,ds$$

$$\geq \left[\int_t^{t^*} p(s)\,ds \right] x(t + \mu) \geq (B/2)x(t + \mu). \qquad (2.1.41)$$

Analogously, integrating over the interval $[t^* - \mu, t]$ we obtain

$$x(t) - x(t^* - \mu) \geq \int_{t^*-\mu}^t p(s)x(s + \mu)\,ds$$

$$\geq \left[\int_{t^*-\mu}^t p(s)\,ds \right] x(t^*) \geq (B/2)x(t^*). \qquad (2.1.42)$$

From (2.1.41) and (2.1.42) it follows that

$$x(t) \geq (B/2)x(t^*) \geq (B/2)^2 x(t + \mu).$$

From the last inequality we obtain that

$$\frac{4}{B^2} \geq \frac{x(t + \mu)}{x(t)} \quad \text{for } t \geq t_2$$

which contradicts equality (2.1.38).

The proofs of (*ii*), (*iii*) and (*iv*) are carried out analogously, and Lemma 2.1.2 is proved.

Let $y(t)$ be a solution of equation (2.1.1). It is immediately verified that the function

$$z(t) = y(t) + P(t)y(t - \tau)$$

is a continuously differentiable solution of the equation

$$z'(t) + R(t)z'(t - \tau) + Q(t)z(t - \sigma) = 0, \quad t \geq t_0 \qquad (2.1.43)$$

where

$$R(t) = P(t - \sigma) \frac{Q(t)}{Q(t - \tau)}.$$

Theorem 2.1.4

Assume conditions H2.1.1, H2.1.3 are fulfilled in addition to condition (2.1.10) and the condition

$$\liminf_{t \to \infty} \int_{t-\tau}^{t-\sigma} \left[-\frac{Q(s - \tau)}{P(s - \sigma)} \right] ds > 1/e. \qquad (2.1.44)$$

Then each solution of equation (2.1.1) oscillates.

Proof

Suppose that this is not true. Let $y(t)$ be an eventually positive solution of equation (2.1.1). Then the function $z(t) = y(t) + P(t)y(t - \tau)$ is a solution of equation (2.1.43). Moreover, from Lemma 2.1.1 (*a*) and (*f*) it follows that

$$z'(t) < 0 \quad \text{and} \quad z(t) < 0. \qquad (2.1.45)$$

Then from (2.1.43) we obtain that

$$\frac{P(t - \sigma)}{Q(t - \tau)} z'(t - \tau) + z(t - \sigma) > 0$$

which implies that

$$z'(t) + \frac{Q(t)}{P(t + \tau - \sigma)} z(t + (\tau - \sigma)) < 0.$$

From Lemma 2.1.2 and (2.1.44) it follows that the last inequality has no eventually negative solutions, which contradicts (2.1.45).

This completes the proof of Theorem 2.1.4.

Example 2.1.6   Consider the equation

$$\frac{d}{dt}[y(t) + (-6 - \sin t)y(t - 2\pi)]$$

$$+ (5 + 2\sin t)y\left(t - \frac{n}{2}\right) = 0, \quad t > 0. \qquad (2.1.46)$$

A straightforward verification yields that all conditions of Theorem 2.1.4 are met. Hence all solutions of equation (2.1.46) oscillate. For instance, $y(t) = \sin t$ is such a solution.

Theorem 2.1.5

Assume conditions H2.1.1, H2.1.2 fulfilled in addition to condition (2.1.4) and the condition

$$\liminf_{t \to \infty} \int_{t-\sigma}^{t} Q(s)\,ds > 1/e. \qquad (2.1.47)$$

Then each solution of equation (2.1.1) oscillates.

Proof

Suppose that this is not true. Let $y(t)$ be an eventually positive solution of equation (2.1.1). Then the function $z(t) = y(t) + P(t)y(t - \tau)$ is a solution of equation (2.1.43). Moreover, from Lemma 2.1.1 (a) and (b) it follows that

$$z'(t) < 0 \quad \text{and} \quad z(t) > 0. \qquad (2.1.48)$$

Then from (2.1.43) we obtain that

$$z'(t) + Q(t)z(t - \sigma) < 0.$$

From Lemma 2.1.2 (iii) and (2.1.47) it follows that the last inequality has no eventually positive solutions which contradicts (2.1.48).
    This completes the proof of Theorem 2.1.5.

Example 2.1.7   Consider the equation

$$\frac{d}{dt}[y(t) + (-\tfrac{1}{2} - \tfrac{1}{8}\cos t)y(t - 2\pi)]$$

$$+ (\tfrac{1}{2} - \tfrac{1}{4}\cos t)y\left(t - \frac{\pi}{2}\right) = 0, \quad t \geq 0. \qquad (2.1.49)$$

A straightforward verification yields that all conditions of Theorem 2.1.5 are met. Hence all solutions of equation (2.1.49) oscillate. For instance, $y(t) = \cos t$ is such a solution.

2.2   First order differential equations with constant coefficients

In this present section a necessary and sufficient condition for oscillation of the solutions of first order neutral differential equations with constant coefficients of the form

$$x'(t) + px'(t - \tau) + qx(t - \sigma) = 0, \quad t \geq t_0 \qquad (2.2.1)$$

is obtained.

Assume the following conditions (H2.2) fulfilled:

H2.2.1 $q, \tau, \sigma = $ const. > 0,

H2.2.2 $p \in \mathbb{R}$.

Our principal hypothesis is that if conditions (H2.2) hold, then the necessary and sufficient condition for all solutions of equation (2.2.1) to oscillate in the sense of Definition 2.1.3 is that the corresponding characteristic equation

$$F(\lambda) \equiv \lambda + p\lambda\,e^{-\lambda\tau} + q\,e^{-\lambda\sigma} = 0 \qquad (2.2.2)$$

should have no real roots.

We shall note that for $p = 0$ the differential equation (2.2.1) takes the form

$$x'(t) + qx(t - \sigma) = 0 \qquad (2.2.3)$$

and the inequality

$$q\sigma > 1/e \qquad (2.2.4)$$

is a necessary and sufficient condition for all solutions of (2.2.3) to oscillate [98]. In this case the characteristic equation is

$$\lambda + q\,e^{-\lambda\sigma} = 0 \qquad\qquad (2.2.5)$$

and (2.2.4) is equivalent to the fact that (2.2.5) has no real roots. For $p = -1$ equation (2.2.1) takes the form

$$x'(t) - x'(t - \tau) + qx(t - \sigma) = 0 \qquad\qquad (2.2.6)$$

and each solution of (2.2.6) oscillates [101]. In this case it is immediately verified that

$$F(\lambda) = \lambda(1 - e^{-\lambda\tau}) + q\,e^{-\lambda\sigma} > 0 \qquad \text{for } \lambda \geq 0,$$

$$F(\lambda) = -\lambda(e^{-\lambda\tau} - 1) + q\,e^{-\lambda\sigma} > 0 \quad \text{for } \lambda \leq 0$$

from which it follows that (2.2.5) has no real roots.

In the proof of Theorem 2.2.1 we shall use the following lemma.

Lemma 2.2.1

Let $x(t)$ be an eventually positive solution of equation (2.2.1). Then the following assertions are valid:

(a) If conditions (H2.2) hold in addition to the condition $p > -1$, then the function $z(t) = x(t) + px(t - \tau)$ is an eventually positive solution of (2.2.1), decreasing and $\lim_{t \to \infty} z(t) = 0$.

(b) If conditions (H2.2) hold in addition to the condition $p < -1$, then the function $z(t) = -x(t) - px(t - \tau)$ is an eventually positive solution of (2.2.1), increasing and $\lim_{t \to \infty} z(t) = +\infty$.

The assertions of Lemma 2.2.1 follow immediately from Lemma 2.1.1.

Theorem 2.2.1

Let conditions (H2.2) hold. A necessary and sufficient condition for all solutions of equation (2.2.1) to oscillate is that the corresponding characteristic equation (2.2.2) should have no real roots.

Proof

Necessity. If $\lambda_0 \in \mathbb{R}$ is a root of equation (2.2.2), then the function $x(t) = e^{\lambda_0 t}$ is a positive solution of the differential equation (2.2.1).

Sufficiency. Suppose that the assertion is not true and not all solutions of (2.2.1) oscillate. Hence there exists a solution $x(t)$ of equation (2.2.1) which is eventually positive.

Consider the following three cases:

(i) *The case* $-1 < p < 0$.

Set

$$z(t) = x(t) + px(t - \tau). \tag{2.2.7}$$

Then by Lemma 2.2.1, $z(t)$ is eventually positive and decreasing and without loss of generality $x(t)$ can be considered also decreasing. It is easy to see that $z(t) < x(t - \sigma)$. Define the set

$$\Lambda(z) = \{\lambda > 0 : z'(t) + \lambda z(t) < 0 \text{ eventually}\}. \tag{2.2.8}$$

From (2.2.1) we have

$$0 = z'(t) + qx(t - \sigma) > z'(t) + qz(t), \text{ eventually}$$

so that $q \in \Lambda(z)$. That is, $\Lambda(z)$ is nonempty.

On the other hand, it follows that

$$0 = z'(t) + qx(t - \sigma) = z'(t) + qz(t - \sigma) - pqx(t - \tau - \sigma)$$

and integrating from t to $t + \tau$ we find

$$z(t + \tau) - z(t) + q \int_{t}^{t+\tau} z(s - \sigma)\,ds - pq \int_{t}^{t+\tau} x(s - \tau - \sigma)\,ds = 0.$$

Taking into account that both $z(t)$ and $x(t)$ are positive and decreasing, the last equation yields

$$z(t + \tau) - z(t) + q\tau z(t + \tau - \sigma) - pq\tau x(t - \sigma) \leqq 0$$

which implies that

$$-pq\tau x(t - \sigma) \leqq z(t).$$

Thus

$$0 = z'(t) + qx(t - \sigma) \leqq z'(t) + \left(\frac{1}{-p\tau}\right) z(t), \text{ eventually}.$$

Therefore $\lambda_0 = 1/(-p\tau)$ is an upper bound of $\Lambda(z)$ which does not depend on z. Thus $\Lambda(z)$ is nonempty and bounded from above.

Let $\lambda \in \Lambda(z)$ and consider the function

$$w(t) \equiv Tz = z(t) + pz(t - \tau). \tag{2.2.9}$$

Set

$$m = \inf_{\lambda > 0} \{ -\lambda - p\lambda\, e^{\lambda\tau} + q\, e^{\lambda\sigma} \} \tag{2.2.10}$$

which is positive because (2.2.2) has no real roots and $\lim_{\lambda \to \infty} (-\lambda - p\lambda e^{\lambda\tau} + q e^{\lambda\sigma}) = +\infty$. We shall show that $\lambda + m \in \Lambda(w)$. From (2.2.1) and (2.2.9) and the fact that both $z(t)$ and $w(t)$ are solutions of (2.2.1) we obtain

$$w'(t) = -qz(t - \sigma). \tag{2.2.11}$$

Define $\varphi(t) = e^{\lambda t}z(t)$. Then

$$\varphi'(t) = [z'(t) + \lambda z(t)]\, e^{\lambda t} < 0, \text{ eventually}$$

and therefore φ is eventually decreasing. Since $z(t) = e^{-\lambda t}\varphi(t)$, (2.2.9) and (2.2.11) give respectively

$$w(t) = e^{-\lambda t}\, \varphi(t) + p\, e^{-\lambda t}\, e^{\lambda\tau}\, \varphi(t - \tau)$$

and

$$w'(t) = -q\, e^{-\lambda t}\, e^{\lambda\sigma}\, \varphi(t - \sigma).$$

Now using the fact that $\varphi(t)$ is decreasing, we obtain

$$
\begin{aligned}
w'(t) &+ (\lambda + m)w(t) \\
&= e^{-\lambda t}[-q\, e^{\lambda\sigma}\, \varphi(t - \sigma) + (\lambda + m)\varphi(t) + (\lambda + m)p\, e^{\lambda\tau}\, \varphi(t - \tau)] \\
&< e^{-\lambda t}\, \varphi(t)[-q\, e^{\lambda\sigma} + \lambda + m + \lambda p\, e^{\lambda\tau} + mp\, e^{\lambda\tau}] \\
&< e^{-\lambda t}\, \varphi(t)[\lambda + p\lambda\, e^{\lambda\tau} - q\, e^{\lambda\sigma} + m] \\
&\leq e^{-\lambda t}\, \varphi(t)[-m + m] = 0
\end{aligned}
\tag{2.2.12}
$$

which implies that $\lambda + m \in \Lambda(w)$. Now set

$$z \equiv z_0, \quad w \equiv Tz_0 = z_1, \quad z_2 = Tz_1 \quad \text{and in general } z_n = Tz_{n-1},$$

$n = 1, 2, \dots$ and observe that for $\lambda \in \Lambda(z) \equiv \Lambda(z_0) \Rightarrow \lambda + nm \in \Lambda(z_n)$, $n = 1, 2, \dots$ which is a contradiction since λ_0 is a common upper bound for all $\Lambda(z_n)$.

(*ii*) *The case p > 0.*
As in case (*i*) set

$$z(t) = x(t) + px(t - \tau) \tag{2.2.7}$$

and

$$\Lambda(z) = \{\lambda > 0: z'(t) + \lambda z(t) < 0 \text{ eventually}\}. \tag{2.2.8}$$

In this case using the characteristic equation of (2.2.1)

$$F(\lambda) = \lambda + \lambda p \, e^{-\lambda \tau} + q \, e^{-\lambda \sigma} = 0$$

we see that if $\tau \geq \sigma > 0$, $F(0) = q > 0$ while $\lim_{\lambda \to -\infty} F(\lambda) = -\infty$ and therefore (2.2.1) always has nonoscillatory solutions. Thus $\sigma > \tau$ is a necessary condition for all solutions of (2.2.1) to oscillate. By Lemma 2.2.1, $z(t)$ is eventually positive and decreasing and without loss of generality $x(t)$ can be considered also decreasing. Thus

$$z(t) = x(t) + px(t - \tau) < x(t - \sigma) + px(t - \sigma) = (1 + p)x(t - \sigma).$$

That is

$$x(t - \sigma) > \frac{1}{1 + p} z(t).$$

From (2.2.1) we have

$$0 = z'(t) + qx(t - \sigma) > z'(t) + \frac{q}{1 + p} z(t), \text{ eventually,}$$

so that

$$\frac{q}{1 + p} \in \Lambda(z).$$

That is, $\Lambda(z)$ is nonempty.
  Now we shall show that $\Lambda(z)$ is bounded from above. Observe that equation (2.2.1) is autonomous and $z(t)$ given by (2.2.7) as a linear combination of solutions of (2.2.1) is itself a solution of (2.2.1) and therefore

$$z'(t) + pz'(t - \tau) + qz(t - \sigma) = 0.$$

Also $z'(t) = -qx(t - \sigma) < 0$ and $z''(t) = -qx'(t - \sigma) > 0$, that is, $z'(t)$ is increasing and therefore $z'(t) > z'(t - \tau)$. Thus the last equation yields

$$(1 + p)z'(t - \tau) + qz(t - \sigma) \leq 0$$

or

$$z'(t) + (q/(1 + p))z(t - (\sigma - \tau)) \leq 0.$$

Integrating the last inequality first from $t - (\sigma - \tau)/2$ to t and then from t to $t + (\sigma - \tau)/2$, we have

$$z(t - (\sigma - \tau)) \leq \left[\frac{2(1 + p)}{q(\sigma - \tau)}\right]^2 z(t). \qquad (2.2.13)$$

Next integrating $z'(t) + qx(t - \sigma) = 0$ over the interval $[t - \sigma + \tau, t]$, we obtain

$$z(t) - z(t - (\sigma - \tau)) + q \int_{t-(\sigma-\tau)}^{t} x(s - \sigma)\,ds = 0$$

and, since $x(t)$ is positive and decreasing and $z(t)$ is positive, we have

$$q(\sigma - \tau)x(t - \sigma) \leq z(t - (\sigma - \tau)). \qquad (2.2.14)$$

Combining (2.2.13) and (2.2.14), we obtain

$$x(t - \sigma) \leq \frac{[2(1 + p)]^2}{q^3(\sigma - \tau)^3} z(t).$$

Thus

$$0 = z'(t) + qx(t - \sigma) \leq z'(t) + \left[\frac{4(1 + p)^2}{q^2(\sigma - \tau)^3}\right]z(t), \text{ eventually.}$$

Therefore

$$\lambda_0 = \frac{4(1 + p)^2}{q^2(\sigma - \tau)^3}$$

is an upper bound of $\Lambda(z)$ which does not depend on z. Thus $\Lambda(z)$ is nonempty and bounded from above.

Now if we follow the same procedure as in the last part of the proof of case (i), considering $\lambda \in \Lambda(z)$ and defining $w(t)$, m as in (2.2.9) and (2.2.10), we shall show that $\lambda + \mu \in \Lambda(w)$, where

$$\mu = \frac{m}{1 + p\, e^{\lambda_0 \tau}} > 0.$$

As in (2.2.12) we obtain

$$w'(\tau) + (\lambda + \mu)w(t)$$
$$= e^{-\lambda t}[-q\, e^{\lambda \sigma}\, \varphi(t - \sigma) + (\lambda + \mu)\varphi(t) + (\lambda + \mu)p\, e^{\lambda \tau}\varphi(t - \tau)]$$

(and since $\varphi(t)$ is decreasing and $\sigma > \tau$)

$$< e^{-\lambda t}\, \varphi(t - \sigma)[-q\, e^{\lambda \sigma} + \lambda + \mu + \lambda p\, e^{\lambda \tau} + \mu p\, e^{\lambda \tau}]$$
$$= e^{-\lambda t}\, \varphi(t - \sigma)[\lambda + p\lambda\, e^{\lambda \tau} - q\, e^{\lambda \sigma} + \mu(1 + p\, e^{\lambda \tau})]$$
$$\leq e^{-\lambda t}\, \varphi(t - \sigma)[-m + \mu(1 + p\, e^{\lambda_0 \tau})] = 0$$

which implies that $\lambda + \mu \in \Lambda(w)$. This (as in case (i)) leads to a contradiction.
(iii) *The case* $p < -1$.
We first assume $\tau \geq \sigma$. Set

$$z(t) = -x(t) - px(t - \tau). \tag{2.2.15}$$

By Lemma 2.2.1, $z(t)$ is eventually positive and increasing and without loss of generality $x(t)$ can be considered also increasing. From (2.2.15) we have

$$-px(t - \tau) > z(t) \Rightarrow x(t - \tau) > \frac{1}{(-p)}\, z(t) \quad \text{or} \quad x(t - \sigma) > \frac{1}{(-p)}\, z(t).$$

Define the set

$$\Lambda(z) = \{\lambda > 0: -z'(t) + \lambda z(t) < 0 \text{ eventually}\}. \tag{2.2.16}$$

From (2.2.1)

$$0 = -z'(t) + qx(t - \sigma) > -z'(t) + \frac{q}{(-p)}\, z(t) \text{ eventually}$$

so that $q/(-p) \in \Lambda(z)$. That is, $\Lambda(z)$ is nonempty.

Now we shall show that $\Lambda(z)$ is bounded from above. From (2.2.15) we have

$$x(t - \tau) = \frac{1}{-p}[x(t) + z(t)]$$

and therefore

$$z'(t) = qx(t - \sigma) = \frac{q}{-p}[x(t+\tau - \sigma) + z(t + \tau - \sigma)].$$

Integrating the last equation from $t - \tau$ to t and taking into account that $x(t)$ and $z(t)$ are eventually positive and increasing, we obtain

$$z(t) - z(t - \tau) = \frac{q}{-p}\int_{t-\tau}^{t}[x(s + \tau - \sigma) + z(s + \tau - \sigma)]\,ds \geq \frac{q}{-p}\tau x(t - \sigma)$$

or

$$x(t - \sigma) \leq \frac{-p}{q\tau}z(t) \text{ eventually.}$$

Thus

$$0 = -z'(t) + qx(t - \sigma) \leq -z'(t) + \left(\frac{-p}{\tau}\right)z(t) \text{ eventually.}$$

Therefore $\lambda_0 = (-p/\tau)$ is an upper bound of $\Lambda(z)$ which does not depend on z. Thus $\Lambda(z)$ is nonempty and bounded from above.

Next we follow a procedure analogous to that given in the last part of the proof of case (i). Let $\lambda \in \Lambda(z)$ and consider the function

$$w(t) = -z(t) - pz(t - \tau) \tag{2.2.17}$$

which is also a solution of (2.2.1) and therefore from (2.2.1)

$$w'(t) = qz(t - \sigma).$$

Set

$$m = \inf_{\lambda > 0}\{\lambda + p\lambda\,e^{-\lambda\tau} + q\,e^{-\lambda\sigma}\} \tag{2.2.18}$$

which by (2.2.2) is positive. We shall show that $\lambda + \mu \in \Lambda(w)$, where

$$\mu = \frac{m}{-p\,e^{-\lambda\tau}} > 0.$$

Define $\varphi(t) = e^{-\lambda t} z(t)$. Then

$$\varphi'(t) = [z'(t) - \lambda z(t)]\,e^{-\lambda t} > 0, \text{ eventually}$$

and therefore φ is eventually increasing. Since $z(t) = e^{\lambda t}\varphi(t)$, similarly as in (2.2.12), we obtain

$$-w'(t) + (\lambda + \mu)w(t)$$
$$= e^{\lambda t}[-q\,e^{-\lambda\sigma}\varphi(t-\sigma) - (\lambda+\mu)\varphi(t) + (\lambda+\mu)(-p)\,e^{-\lambda\tau}\varphi(t-\tau)]$$

(and since $\varphi(t)$ is increasing and $\tau \geq \sigma$)

$$< e^{\lambda t}\varphi(t-\sigma)[-q\,e^{-\lambda\sigma} - (\lambda+\mu) + (\lambda+\mu)(-p)\,e^{-\lambda\tau}]$$
$$= e^{\lambda t}\varphi(t-\sigma)[-\lambda - p\lambda\,e^{-\lambda\tau} - q\,e^{-\lambda\sigma} + \mu(-p\,e^{-\lambda\tau} - 1)]$$
$$\leq e^{\lambda t}\varphi(t-\sigma)[-m + \mu(-p)\,e^{-\lambda\tau} - \mu]$$
$$< e^{\lambda t}\varphi(t-\sigma)[-m + \mu(-p)\,e^{-\lambda_0\tau}] = 0$$

which implies that $\lambda + \mu \in \Lambda(w)$ and, as before, we are led to a contradiction.

To complete the proof in this case that $p < -1$ we have to assume that $\sigma > \tau$. Here we set

$$z(t) = -x(t) - px(t-\tau) + q\int_{t-\sigma}^{t-\tau} x(s)\,ds \qquad (2.2.19)$$

which is a solution of (2.2.1) and which is eventually positive. From (2.2.1) we have

$$z'(t) = qx(t-\tau) > 0$$

which implies that $z(t)$ is increasing and without loss of generality $x(t)$ can be considered also increasing. Therefore (2.2.19) yields

$$z(t) < -px(t-\tau) + q\int_{t-\sigma}^{t-\tau} x(s)\,ds < -px(t-\tau) + q(\sigma-\tau)x(t-\tau)$$

$$= [q(\sigma-\tau) - p]x(t-\tau)$$

or

$$x(t - \tau) > \frac{1}{q(\sigma - \tau) - p} z(t).$$

As before, we define the set

$$\Lambda(z) = \{\lambda > 0: -z'(t) + \lambda z(t) < 0 \text{ eventually}\}. \qquad (2.2.16)$$

In view of the last inequality, we have

$$0 = -z'(t) + qx(t - \tau) > -z'(t) + \frac{q}{q(\sigma - \tau) - p} z(t), \text{ eventually}$$

so that

$$\frac{q}{q(\sigma - \tau) - p} \in \Lambda(z).$$

That is, $\Lambda(z)$ is nonempty.

Next we shall show that $\Lambda(z)$ is bounded from above. Since $z(t)$ is a solution of (2.2.1), it follows that

$$z'(t) + pz'(t - \tau) + qz(t - \sigma) = 0$$

which implies

$$z'(t) + pz'(t - \tau) \leqq 0$$

But $z'(t) = qx(t - \tau)$ and thus

$$qx(t - \tau) + pz'(t - \tau) \leqq 0.$$

Integrating the last inequality in the interval $[t, t + \tau]$ and taking into account that $x(t)$ is positive and increasing and $z(t)$ eventually positive, we obtain

$$0 \geqq q \int_{t}^{t+\tau} x(s - \tau) \, ds + pz(t) - pz(t - \tau)$$

$$\geqq q\tau x(t - \tau) + pz(t) - pz(t - \tau)$$

$$> q\tau x(t - \tau) + pz(t)$$

or

$$x(t - \tau) \leq \frac{(-p)}{q\tau} z(t), \text{ eventually.}$$

Thus

$$0 = -z'(t) + qx(t - \tau) \leqq -z'(t) + (-p/\tau)z(t), \text{ eventually,}$$

which implies that $\lambda_0 = (-p/\tau)$ is an upper bound of $\Lambda(z)$ which does not depend on z. Thus $\Lambda(z)$ is nonempty and bounded from above.

Now let $\lambda \in \Lambda(z)$ and consider the function (cf. (2.2.19))

$$w(t) = -z(t) - pz(t - \tau) + q \int_{t-\sigma}^{t-\tau} z(s)\, ds.$$

Using the fact that $z(t)$ and $w(t)$ are solutions of (2.2.1), we obtain

$$w'(t) = qz(t - \tau).$$

We define m as in (2.2.18) and $\varphi(t) = e^{-\lambda t} z(t)$. We shall show that $\lambda + \mu \in \Lambda(w)$, where

$$\mu = \frac{m\, e^{\lambda_0 \tau}}{q\lambda_0^{-1} + (-p)} > 0.$$

We have

$$-w'(t) + (\lambda + \mu)w(t)$$

$$= e^{\lambda t}\Bigg[-q\, e^{-\lambda \tau}\, \varphi(t - \tau) - (\lambda + \mu)\varphi(t) + (\lambda + \mu)(-p)\, e^{-\lambda \tau}\, \varphi(t - \tau)$$

$$+ (\lambda + \mu)q\, e^{-\lambda t} \int_{t-\sigma}^{t-\tau} e^{\lambda s}\, \varphi(s)\, ds \Bigg]$$

(and since $\varphi(t)$ is increasing and $\sigma > \tau$)

$$< e^{\lambda t} \varphi(t - \tau) \left[-q\,e^{-\lambda \tau} - (\lambda + \mu) + (\lambda + \mu)(-p)\,e^{-\lambda \tau} + (\lambda + \mu)q\,\frac{e^{-\lambda t}}{\lambda} \right.$$

$$\left. \times\, (e^{\lambda(t - \tau)} - e^{\lambda(t - \sigma)}) \right]$$

$$= e^{\lambda t} \varphi(t - \tau) \left[-q\,e^{-\lambda \tau} - \lambda - \mu - \lambda p\,e^{-\lambda \tau} - \mu p\,e^{-\lambda \tau} + q\,e^{-\lambda \tau} - q\,e^{-\lambda \sigma} \right.$$

$$\left. + \frac{\mu q}{\lambda}\,e^{-\lambda \tau} - \frac{\mu q}{\lambda}\,e^{-\lambda \sigma} \right]$$

$$< e^{\lambda t} \varphi(t - \tau) \left[-\lambda - p\lambda\,e^{-\lambda \tau} - q\,e^{-\lambda \sigma} - \mu p\,e^{-\lambda \tau} + \frac{\mu q}{\lambda}\,e^{-\lambda \tau} \right]$$

$$\leqq e^{\lambda t} \varphi(t - \tau) \left[-m + \mu\,e^{-\lambda_0 \tau} \left(\frac{q}{\lambda_0} + (-p) \right) \right] = 0$$

which implies that $\lambda + \mu \in \Lambda(w)$ and we are led to a contradiction. The proof of Theorem 2.2.1 is complete.

Corollary 2.2.1   *Let conditions (H2.2) hold,*

$$-1 < p < 0 \tag{2.2.17}$$

and

$$(-p)\tau q > 1. \tag{2.2.18}$$

Then all solutions of equation (2.2.1) oscillate.

Proof

Suppose that this is not true, that is, that not all solutions of (2.2.1) oscillate. Hence there exists a solution $x(t)$ of equation (2.2.1) which is eventually positive. Then from the proof of Theorem 2.2.1 we obtain that $q \in \Lambda(z)$ and the number $1/(-p\tau)$ is an upper bound for the set $\Lambda(z)$. Hence $q \leqq 1/(-p\tau)$ which contradicts condition (2.2.18).

Analogously, using Theorem 2.2.1, the following corollaries are proved.

Corollary 2.2.2 *Let conditions (H2.2) hold,*

$$p > 0 \qquad (2.2.19)$$

and

$$\left(\frac{q}{1+p}\right)^3 \frac{(\sigma - \tau)^3}{4} > 1. \qquad (2.2.20)$$

Then all solutions of equation (2.2.1) oscillate.

Corollary 2.2.3 *Let conditions (H2.2) hold,*

$$p < -1 \qquad (2.2.21)$$

$$\tau \geqq \sigma \qquad (2.2.22)$$

and

$$q\tau/p^2 > 1 \qquad (2.2.23)$$

Then all solutions of equation (2.2.1) oscillate.

Corollary 2.2.4 *Let conditions (H2.2) hold, condition (2.2.21),*

$$\tau < \sigma \qquad (2.2.24)$$

and

$$\frac{q\tau}{p^2 + (-p)q(\sigma - \tau)} > 1. \qquad (2.2.25)$$

Then all solutions of equation (2.2.1) oscillate.

2.3 First order differential equations with distributed delay

The oscillatory and asymptotic properties of the solutions of functional differential equations of retarding type have been an object of investigation reported in many works. An extensive bibliography on this subject is given in [141]. We shall note that analogous results obtained for neutral equations are quite few. Sufficient conditions for oscillation of the solutions of neutral equations in the case of one discrete delay are obtained in the works [160], [7], [15], [125].

In the work [139] it is proved that for the oscillation of all solutions of the equation

$$x'(t) + px'(t - \tau) + qx(t - \sigma) = 0, \quad t \geq t_0, \quad \tau, \sigma > 0$$

it is necessary and sufficient that the corresponding characteristic equation should have no real roots. In the present section analogous results are obtained for a considerably wider class of equations of the form

$$\frac{d}{dt}\left[x(t) + \delta_1 \int_0^{\tau_1} x(t - s) dr_1(s) \right] + \delta_2 \int_0^{\tau_2} x(t - s) dr_2(s) = 0 \qquad (2.3.1)$$

where $\delta_1, \delta_2 = \pm 1$.

We shall say that conditions (H2.3) are met if the following conditions hold:

H2.3.1   The functions $r_i : [0, \tau_i] \to [0, \infty)$, $i = 1, 2$ are nondecreasing.

H2.3.2 $\tau_i > 0$, $r_i(0) = 0$, $i = 1, 2$, $r_1(0^+) = 0$ and $r_2(\tau_2) > 0$.

Our main conjecture is that if conditions (H2.3) hold, for the oscillation of all nonzero solutions of equation (2.3.1) it is necessary and sufficient that the corresponding characteristic equation

$$Q(z) = z\left[1 + \delta_1 \int_0^{\tau_1} e^{-zs} dr_1(s) \right] + \delta_2 \int_0^{\tau_2} e^{-zs} dr_2(s) = 0 \qquad (2.3.2)$$

should have no real roots.

The necessity of this condition is obvious since if $z^* \in \mathbb{R}$ is a root of equation (2.3.2), then the function $x(t) = e^{z^* t}$ is a nonoscillatory solution of equation (2.3.1). Hence, in particular, it follows that our main conjecture holds in the following three cases:

(i) $\delta_2 = -1$

(ii) $\delta_1 = \delta_2 = 1$, $\tau_2 > \tau_1$, $r_1(\tau_1) > r_1(\tau_2)$

(iii) $\delta_1 = \delta_2 = 1$, $\tau_1 = \tau_2$, $r_1(\tau_1) > r_1(\tau_1^-)$

since in each one of cases (i)–(iii) equation (2.3.2) has at least one real root.

In view of the above arguments we shall assume without a special stipulation that $\delta_2 = 1$.

Definition 2.3.1 The function $f:[t_f, \infty) \to \mathbb{R}$ is said to oscillate if there exists a strictly increasing sequence

$$\{t_i\}_{i=1}^{\infty} \subset [t_f, \infty), \quad \lim_{i \to \infty} t_i = \infty \text{ such that } f(t_i)f(t_{i+1}) < 0.$$

Definition 2.3.2 The function $f:[t_f, \infty) \to \mathbb{R}$ is said to be monotone if it is nondecreasing or nonincreasing.

Definition 2.3.3 The function $f:[t_f, \infty) \to \mathbb{R}$ is said to be vanishing if eventually $f(t) \equiv 0$.

Definition 2.3.4 The continuous function $x:[t_x, \infty) \to \mathbb{R}$ is said to be a solution of equation (2.3.1) and we shall write $x \in (2.3.1)$ if the function $x(t) + \delta_1 \int_0^{\tau_1} x(t-s)\, dr_1(s)$ is continuously differentiable for $t \geq t_x + \tau_1$ and equation (2.3.1) is satisfied by x for $t \geq t_x + \tau$, $\tau = \max\{\tau_1, \tau_2\}$.

Lemma 2.3.1

Let conditions (H2.3) hold, let the function $r:[0, \sigma] \to \mathbb{R}$ be of bounded variation and let $x:[t_x, \infty) \to \mathbb{R}$ be a solution of equation (2.3.1).
Then the function $y:[t_y, \infty) \to \mathbb{R}$, $y(t) = \int_0^{\sigma} x(t-s)\, dr(s)$, $t_y = t_x + \sigma$ is also a solution of equation (2.3.1).

Proof

Since $y(t)$ is a continuous function, if we denote

$$y_1(t) = x(t) + \delta_1 \int_0^{\tau_1} x(t-s)\, dr_1(s), \quad t \geq t_x + \tau_1$$

and change the order of integration, we obtain for $t \geq t_y + \tau_1$ the equalities

$$y(t) + \delta_1 \int_0^{\tau_1} y(t-s)\, dr_1(s) = \int_0^{\sigma} x(t-s)\, dr(s) + \int_0^{\tau_1} \left[\int_0^{\sigma} x(t-s-h)\, dr(h) \right] dr_1(s)$$

$$= \int_0^{\sigma} y_1(t-h)\, dr(h).$$

The above equality implies that the function $y(t) + \delta \int_0^{\tau_1} y(t - s)\, dr_1(s)$ is continuously differentiable with respect to t with derivative

$$\int_0^\sigma y_1'(t - h)\, dr(h) = \int_0^\sigma \left[\int_0^{\tau_2} x(t - h - s)\, dr_2(s)\right] dr(h)$$

$$= -\int_0^{\tau_2} \left[\int_0^\sigma x(t - s - h)\, dr(h)\right] dr_2(s)$$

$$= -\int_0^{\tau_2} y(t - s)\, dr_2(s).$$

This completes the proof of Lemma 2.3.1.

Lemma 2.3.2

Let conditions (H2.3) *hold. Then if the function* $x \in (2.3.1)$ *is bounded, then* $\lim_{t \to \infty} \inf |x(t)| = 0.$

Proof

Let $\lim_{t \to \infty} \inf |x(t)| = c > 0$. Then, integrating both sides of equation (2.3.1) from $t_0 \geq t_x + \tau$ to $t > t_0$, we obtain the equality

$$x(t) + \delta_1 \int_0^{\tau_1} x(t - s)\, dr_1(s) - \left[x(t_0) + \delta_1 \int_0^{\tau_1} x(t_0 - s)\, dr_1(s)\right]$$

$$= -\int_{t_0}^t \left[\int_0^{\tau_2} x(t' - s)\, dr_2(s)\right] dt'.$$

For $t \to \infty$ the left-hand side of the equality is bounded and the modulus of its right-hand side is not smaller than the expression $c_2 r_2(\tau_2) t + o(t)$ which is impossible.

Lemma 2.3.2 is proved.

Corollary 2.3.1   *If conditions* (H2.3) *hold and the function* $x \in (2.3.1)$ *is monotone, then* $x(\infty) \in \{0, -\infty, \infty\}$.

Remark 2.3.1 Lemma 2.3.2 and Corollary 2.3.1 are immediately generalized for a considerably wider class of equations of the form

$$\frac{d}{dt}\left[x(t) + \int_0^{\tau_1(t)} x(t-s)\, d_s r_1(t,s) \right] + \int_0^{\tau_2(t)} x(t-s)\, d_s r_2(t,s) = 0$$

where

$$\tau_i(t) > 0, \quad \lim_{t \to \infty} (t - \tau_i(t)) = \infty, \quad i = 1, 2,$$

the function $r_1(t,s)$ is of uniformly bounded variation with respect to s, the function $r_2(t,s)$ is monotone with respect to s and $\inf_t |r_2(t,\tau_2(t)) - r_2(t,0)| > 0$.
Let $x \in (2.3.1)$ and introduce the notation

$$y_0(t) = x(t), \quad y_i(t) = y_{i-1}(t) + \delta_1 \int_0^{\tau_1} y_{i-1}(t-s)\, dr_1(s), \quad i = 1, 2, \ldots$$

$$(2.3.3)$$

By induction on i and by Lemma 2.3.1 we deduce that each function $y_i(t)$, $i = 1, 2, \ldots$ is defined in the interval $[t_x + i\tau_1, \infty)$, continuously differentiable in it and it is a solution of equation (2.3.1). Moreover, if $x(t)$ is a nonoscillating function, then from equation (2.3.1) by induction on i we obtain that the functions $y_i(t)$, $i = 1, 2, \ldots$ are eventually monotone and by Corollary 2.3.1 $y_i(\infty) \in \{0, -\infty, \infty\}$.

Lemma 2.3.3

Let conditions ($H2.3$) hold in addition to one of the following two conditions:
  (iv) $\delta_1 = 1$.
  (v) $\delta_1 = -1$, $r_1(\tau_1) < 1$.
  Then if the function $x \in (2.3.1)$ is nonnegative and nonvanishing (positive),
then all functions $y_i(t)$, $i = 1, 2, \ldots$ are eventually positive and eventually nonincreasing (decreasing).

Proof

It suffices to consider only the case $i = 1$ and after that the proof is completed by induction on i.

If condition (*iv*) holds, then all assertions of Lemma 2.3.3 follow immediately from equation (2.3.1). Provided that condition (*v*) holds, it suffices to show that $y_1(\infty) = 0$ and then the assertions needed follow immediately from equation (2.3.1). In fact, if we suppose that $y_1(\infty) = -\infty$, then in view of $y_1 \in (2.3.1)$ from equation (2.3.1) we deduce that $y_2'(\infty) = \infty$, hence eventually the following inequality holds

$$y_1(t) - \int_0^{\tau_1} y_1(t-s)\,dr_1(s) = y_2(t) > 0.$$

Since the function $y_1(t)$ is eventually nonincreasing, then eventually the inequality

$$y_1(t) - \int_0^{\tau_1} y_1(t-s)\,dr_1(s) \le y_1(t)[1 - r_1(\tau_1)] < 0$$

holds which is impossible. This completes the proof of Lemma 2.3.3.

Lemma 2.3.4

Let conditions ($H2.3$) hold and $\delta_1 = -1$ and $r_1(\tau_1) \ge 1$.
    Then if the function $x \in (2.3.1)$ is nonnegative and nonvanishing, then $y_i(\infty) = (-1)^i \infty$, $i = 1, 2, \ldots$.

Proof

It is immediately seen that it suffices to prove that $y_1(\infty) = -\infty$.
    If we assume that $y_1(\infty) = 0$, then $y_2(\infty) = 0$ as well and since $x(t)$ is eventually nonnegative and $y_1(t)$ and $y_2(t) \in (2.3.1)$, then we deduce that the functions $y_1(t)$ and $y_2(t)$ are eventually positive and nonincreasing which implies that eventually the following inequalities hold

$$y_1(t) > \int_0^{\tau_1} y_1(t-s)\,dr_1(s) \ge \int_0^{\tau_1} y_1(t)\,dr_1(s) = r_1(\tau_1)y_1(t)$$

which is impossible. Thus Lemma 2.3.4 is proved.

Corollary 2.3.2 *Let the conditions of Lemma 2.3.3 (Lemma 2.3.4) hold.*
Then if equation (2.3.1) has a nonnegative nonvanishing solutions, then it has a positive decreasing (increasing) solution.

Theorem 2.3.1

Let conditions (H2.3) hold and $\delta_1 = -1$ and $r_1(\tau_1) < 1$.
Then for all nonvanishing solutions of equation (2.3.1) to oscillate it is necessary and sufficient that equation (2.3.2) should have no real roots.

Proof

The necessity of the condition that equation (2.3.2) should have no real roots has been already considered.

Sufficiency. Suppose that equation (2.3.2) has no real roots and equation (2.3.1) has a nonnegative nonvanishing solution $x(t)$.

Then by Corollary 2.3.2 and Lemma 2.3.3 the functions $y_i(t)$, $i = 1, 2, \ldots$ are eventually positive and eventually decreasing. On the other hand, from the fact that the quasipolynomial $Q(z)$ has no real roots it follows that at least one of the functions $r_i(t)$, $i = 1, 2$, is not constant for $t > 0$, hence $Q(-\infty) = \infty$, from which we deduce that the following inequality holds

$$q = \min_{z \in [0, \infty)} Q(z) > 0.$$

Introduce the sets

$$Z_i = \{z \in [0, \infty): y_i'(t) + z y_i(t) \leq 0 \text{ eventually}\}, \quad i = 1, 2, \ldots \quad (2.3.4)$$

Since $0 \in Z_i$, $i = 1, 2, \ldots$, then the set $Z_i \neq \varnothing$ for any $i = 1, 2, \ldots$. On the other hand, if $z \in Z_i$, then we can conclude that $y_i(t) = O(e^{-zt})$ as $t \to \infty$. In view of Henry's theorem [68] which states that a nonvanishing solution of equation (2.3.1) cannot tend to zero as $t \to \infty$ faster than any exponential, we conclude that each set Z_i is bounded.

Choose $z^* \in Z_1$ such that $z^* > \sup Z_1 - q$ and set $\varphi(t) = e^{z^* t} y_1(t)$. Relation (2.3.4) implies that eventually $\varphi'(t) \leq 0$. From this inequality, the definition of the constant q and the fact that $y_1(t)$ is a solution of equation

(2.3.1) it follows that eventually the following estimate holds

$$y_1'(t) + (z^* + q)y_1(t)$$

$$= \int_0^{\tau_1} (e^{-z^*(t-s)} \varphi(t-s))' \, dr_1(s) - \int_0^{\tau_2} e^{-z^*(t-s)} \varphi(t-s) \, dr_2(s)$$

$$+ (z^* + q) e^{-z^*t} \varphi(t)$$

$$= -z^* \int_0^{\tau_1} e^{-z^*(t-s)} \varphi(t-s) \, dr_1(s) + \int_0^{\tau_1} e^{-z^*(t-s)} \varphi'(t-s) \, dr_1(s)$$

$$- \int_0^{\tau_2} e^{-z^*(t-s)} \varphi(t-s) \, dr_2(s) + (z^* + q) e^{-z^*t} \varphi(t)$$

$$\leqq e^{-z^*t} \varphi(t) \left[z^* - z^* \int_0^{\tau_1} e^{z^*s} \, dr_1(s) - \int_0^{\tau_2} e^{z^*s} \, dr_2(s) + q \right]$$

$$= e^{-z^*t} \varphi(t)[q - Q(-z^*)] < 0.$$

Hence $z^* + q \in Z_1$ which contradicts the choice of z^*.
This completes the proof of Theorem 2.3.1.

Theorem 2.3.2

Suppose the following conditions are fulfilled:
(1)   Conditions (H2.3) hold.
(2) $\delta_1 = 1$, $\tau_2 > \tau_1$ and $r_2(\tau_2) > r_2(\tau_1^+)$.
(3)   The equation

$$\tilde{Q}(z) = z \left(1 + \int_0^{\tau_1} e^{-zs} \, dr_1(s) \right) + \int_{\tau_1}^{\tau_2} e^{-zs} \, dr_2(s) = 0$$

has no real roots.
  Then each nonvanishing solution of equation (2.3.1) oscillates.

Proof

Suppose that equation (2.3.1) has a nonvanishing solution $x : [t_x, \infty) \to \mathbb{R}$ which is eventually with constant signs. Without loss of generality in view of Corollary 2.3.2 we can assume that the function $x(t)$ is eventually positive and decreasing.

Then from equality (2.3.3) and equation (2.3.1) we deduce that the functions $y_i(t)$, $i = 1, 2, \ldots$ are positive, decreasing and convex for $t \geq t_i = t_x + i\tau_2$.

If we choose the number $\tau_3 \in (\tau_1, \tau_2)$ so that the inequality $b = r_2(\tau_2) - r_2(\tau_3) > 0$ holds, then equation (2.3.1) for $t \geq t_i + \tau_2$ implies the inequality

$$y_i'(t - \tau_1)(1 + r_1(\tau_1)) + by_i(t - \tau_3) \leq 0.$$

Integrating the last inequality from $t + \tau_1 \geq t_i + \tau_2$ to $t + 2\tau_1$, we obtain that for $t \geq t_{i+1} - \tau_1$ the following inequalities hold

$$y_i(t)(1 + r_1(\tau_1)) \geq y_i(t + \tau_1)(1 + r_1(\tau_1)) + b \int_{t+\tau_1}^{t+2\tau_1} y_i(s - \tau_3)\, ds$$

$$\geq b\tau_1 y_i(t - (\tau_3 - \tau_1)).$$

From the last inequalities, choosing $c_i > 0$ small enough, we deduce for $t \geq t_i$ the inequality

$$y_i(t) \geq c_i \left(\frac{b\tau_1}{1 + r_1(\tau_1)}\right)^{t/(\tau_3 - \tau_1)}$$

which implies that all sets Z_i satisfy the estimate

$$\sup_z Z_i \leq z_0 = \frac{1}{\tau_3 - \tau_1} \ln \frac{1 + r_1(\tau_1)}{b\tau_1}, \quad i = 1, 2, \ldots$$

$(Z_i \neq \varnothing$ since $0 \in Z_i$, $i = 1, 2, \ldots)$. In view of $\tilde{Q}(-\infty) = \infty$, setting

$$h = \left[1 + \int_0^{\tau_1} e^{z_0 s}\, dr_1(s)\right]^{-1} \min_{z \in [0, \infty)} \tilde{Q}(-z)$$

we conclude that $h > 0$.

Let $z^* \in Z_i$ for some $i \geq 1$ and set $\varphi_i(t) = y_i(t) e^{z^* t}$. Then the function $\varphi'(t) \leq 0$ eventually and the following inequalities hold

$$y'_{i+1}(t) + (z^* + h) y_{i+1}(t)$$

$$= - \int_0^{\tau_2} e^{-z^*(t-s)} \varphi_i(t-s) \, dr_2(s) + (z^* + h) e^{-z^* t} \varphi_i(t)$$

$$+ (z^* + h) \int_0^{\tau_1} e^{-z^*(t-s)} \varphi_i(t-s) \, dr_1(s)$$

$$\leq e^{-z^* t} \varphi(t - \tau_1) \left[- \int_0^{\tau_2} e^{z^* s} \, dr_2(s) + z^* + z^* \int_0^{\tau_1} e^{z^* s} \, dr_1(s) \right.$$

$$\left. + h \left(1 + \int_0^{\tau_1} e^{z_0 s} \, dr_1(s) \right) \right]$$

$$= e^{-z^* t} \varphi_i(t - \tau_1) \left[\min_{z \in [0, \infty)} \tilde{Q}(-z) - \tilde{Q}(-z^*) \right] \leq 0.$$

Hence $z^* + h \in Z_{i+1}$ which implies that for each $i \geq 1$ the inequality $\sup_z Z_{i+1} \geq \sup_z Z_i + h$ holds which is impossible since

$$\sup_i \sup_z Z_i \leq z_0.$$

This completes the proof of Theorem 2.3.2.

Theorem 2.3.3

Let conditions (H2.3) hold, $\delta_1 = -1$, $\tau_2 < \tau_1$, $r_1(\tau_2) = 0$ and $r_1(\tau_1) > 1$.
  Then for all nonvanishing solutions of equation (2.3.1) to be nonoscillating it is necessary and sufficient that equation (2.3.2) should have no real roots.

Proof

Suppose that the quasipolynomial $Q(z)$ has no real roots and equation (2.3.1) has a positive strictly increasing solution $x(t)$.

Introduce the functions

$$\bar{y}_i(t) = (-1)^i y_i(t), \quad i = 0, 1, \ldots, \quad t \geq t_x + i\tau_1.$$

A straightforward verification yields that they satisfy the relations

$$\bar{y}_i(t) = \int_0^{\tau_1} \bar{y}_{i-1}(t - s) \, dr_1(s) - \bar{y}_{i-1}(t),$$

$$\bar{y}_i'(t) = \int_0^{\tau_2} \bar{y}_{i-1}(t - s) \, dr_2(s), \quad i \geq 1.$$

which imply that all functions $\bar{y}_i(t)$, $i = 1, 2, \ldots$ are eventually positive, eventually increasing and eventually convex.

Introduce the sets

$$\bar{Z}_i = \{z \in [0, \infty) : \bar{y}_i'(t) - z\bar{y}_i(t) \geq 0 \text{ eventually}\}, \quad i = 1, 2, \ldots$$

It is immediately seen that $0 \in \bar{Z}_i$, hence $\bar{Z}_i \neq \varnothing$ for $i = 1, 2, \ldots$. On the other hand, equation (2.3.1) implies that eventually the following inequalities hold

$$0 > \bar{y}_i'(t) - \int_{\tau_2}^{\tau_1} \bar{y}_i'(t - s) \, dr_1(s) > \bar{y}_i'(t) - r_1(\tau_1)\bar{y}_i'(t - \tau_2) \quad (2.3.5)$$

From inequalities (2.3.5) we deduce $\bar{y}_i' = O([r_1(\tau_1)]^{t/\tau_2})$ as $t \to \infty$, from which we conclude that $\bar{y}_i(t) = O([r_1(\tau_1)]^{t/\tau_2})$ as $t \to \infty$, hence

$$\sup_z \bar{Z}_i \leq z_0 = \frac{1}{\tau_2} \ln r_1(\tau_1), \quad i = 1, 2, \ldots \quad .$$

In view of $Q(\infty) = \infty$, setting

$$h = \frac{1}{r_1(\tau_1)} \cdot \min_{z \in [0, \infty)} Q(z)$$

we conclude that $h > 0$.

Let $z^* \in \bar{Z}_i$ for some $i \geq 1$ and set $\varphi_i(t) = \bar{y}_i(t) e^{-z^* t}$. Then the function $\varphi_i'(t) \geq 0$ eventually and the following inequalities hold

$$\bar{y}_{i+1}'(t) - (z^* + h)\bar{y}_{i+1}(t)$$

$$= \int_0^{\tau_2} e^{z^*(t-s)} \varphi_i(t-s) \, dr_2(s) - (z^* + h)$$

$$\times \left[\int_{\tau_2}^{\tau_1} e^{z^*(t-s)} \varphi_i(t-s) \, dr_1(s) - e^{z^* t} \varphi_i(t) \right]$$

$$\geq e^{z^* t} \varphi_i(t - \tau_2) \int_0^{\tau_2} e^{-z^* s} \, dr_2(s) - (z^* + h) e^{z^* t} \varphi_i(t - \tau_2)$$

$$\times \left[\int_{\tau_2}^{\tau_1} e^{-z^* s} \, dr_1(s) - 1 \right]$$

$$\geq e^{z^* t} \varphi_i(t - \tau_2) \left[\int_0^{\tau_2} e^{-z^* s} \, dr_2(s) + z^* - z^* \int_{\tau_2}^{\tau_1} e^{-z^* s} \, dr_1(s) - hr_1(\tau_1) \right]$$

$$= e^{z^* t} \varphi_i(t - \tau_2)[Q(z^*) - hr_1(\tau_1)] \geq 0.$$

Hence $z^* + h \in \bar{Z}_{i+1}$, from which it follows that for each $i \geq 1$ the inequality $\sup_z \bar{Z}_{i+1} \geq \sup_z \bar{Z}_i + h$ holds which is impossible since

$$\sup_i \sup_z \bar{Z}_i \leq z_0.$$

Theorem 2.3.3 is proved.

Theorem 2.3.4

Let conditions $(H2.3)$ hold, $\delta_1 = -1$, $\tau_2 < \tau_1$, $r_1(\tau_2) = 0$ and $r_1(\tau_1) = 1$. Then each nonvanishing solution of equation $(2.3.1)$ oscillates.

Proof

As in the proof of Theorem 2.3.3 we conclude that eventually inequalities $(2.3.5)$ hold, which contradicts the eventual convexity of the functions $\bar{y}_i(t)$.

This completes the proof of Theorem 2.3.4.

Corollary 2.3.3 *Let conditions ($H2.3$) hold, $\delta_1 = -1, \tau_2 < \tau_1$ and $r_1(\tau_2) = 0$. Then the main conjecture is valid.*

The proof of Corollary 2.3.3 follows from Theorems 2.3.1, 2.3.3 and 2.3.4.

2.4   First order nonlinear differential equations

In this section the oscillatory properties are investigated and conditions for existence of bounded nonoscillating solutions are obtained for first order neutral differential equations of the form

$$\frac{d}{dt}[x(t) + h(t)x(\tau(t))] = f(t, x(g_1(t)), \dots, x(g_N(t))) \qquad (2.4.1)$$

and

$$\frac{d}{dt}[x(t) + h(t)x(\tau(t))] + f(t, x(g_1(t)), \dots, x(g_N(t))) = 0 \qquad (2.4.2)$$

We shall say that conditions (H2.4) are met if the following conditions hold:

H2.4.1 $h(t) \in C([t_0, \infty); (0, \infty))$.

H2.4.2 $\tau(t) \in C([t_0, \infty); \mathbb{R})$,
$\tau(t)$ is a strictly increasing function and $\lim_{t \to \infty} \tau(t) = \infty$.

H2.4.3 $g_i(t) \in C([t_0, \infty); \mathbb{R})$,

$$\lim_{t \to \infty} g_i(t) = \infty, \quad i = 1, 2, \dots, N.$$

H2.4.4 $f(t, y_1, \dots, y_N) \in C([t_0, \infty) \times \mathbb{R}^N; \mathbb{R})$,
$f(t, y_1, \dots, y_N)$ is a nondecreasing function with respect to each variable y_i, $i = 1, 2, \dots, N$ and

$$y_1 f(t, y_1, \dots, y_N) \geqq 0 \text{ for } y_1 y_i > 0, \quad i = 1, 2, \dots, N.$$

Definition 2.4.1   The function $x(t) : [T_x, \infty) \to \mathbb{R}$ is said to be a solution of equation (2.4.1) ((2.4.2)) if $x(t) + h(t)x(\tau(t))$ is a continuously differentiable function and $x(t)$ satisfies for $t \geqq T_x$ the differential equation (2.4.1) ((2.4.2)), where T_x is a sufficiently large number.

Definition 2.4.2   The solution $x(t)$ of equation $(2.4.1)$ $((2.4.2))$ which does not vanish eventually is said to oscillate if the function $x(t)$ has a sequence of zeros tending to $+\infty$. Otherwise, the solution is said to be nonoscillating.

Definition 2.4.3   The function $f(t, y_1, \ldots, y_N)$ is said to be strongly superlinear if there exists constant $\alpha > 1$ and $M > 0$ such that the function $|z|^{-\alpha} |f(t, z, \ldots, z)|$ is nondecreasing with respect to $|z|$ for $|z| \geq M$.

Definition 2.4.4   The function $f(t, y_1, \ldots, y_N)$ is said to be strongly sublinear if there exist constants $0 < \beta < 1$ and $m > 0$ such that the function $|z|^{-\beta} |f(t, z, \ldots, z)|$ is nonincreasing with respect to $|z|$ for $0 < |z| \leq m$.

Moreover, we assume that together with conditions (H2.4) one of the following four conditions is met:

   H2.4.5 $h(t) \leq \lambda < 1, \quad \tau(t) < t,$
   H2.4.6 $1 < \mu \leq h(t) \leq v, \quad \tau(t) > t,$
   H2.4.7 $h(t) \leq \lambda < 1, \quad \tau(t) > t,$
   H2.4.8 $1 < \mu \leq h(t) \leq v, \quad \tau(t) < t,$ where $\lambda, \mu, v = $ const.
   Introduce the following notation:

$$\tau^0(t) \equiv t, \quad \tau^i(t) = \tau(\tau^{i-1}(t)),$$

$$\tau^{-i}(t) = \tau^{-1}(\tau^{-(i-1)}(t)), \quad i = 1, 2, \ldots \tag{2.4.3}$$

where $\tau^{-1}(t)$ stands for the inverse function of $\tau(t)$,

$$A(g_*) = \{t \in [t_0, \infty) : g_*(t) > t\},$$

$$A(g_*, \tau) = \{t \in [t_0, \infty) : g_*(t) > \tau(t) \geq t_0\},$$

$$R(g^*) = \{t \in [t_0, \infty) : t_0 \leq g^*(t) < t\}, \tag{2.4.4}$$

$$R(g^*, \tau) = \{t \in [t_0, \infty) : t_0 \leq g^*(t) < \tau(t)\}$$

where

$$g_*(t) = \min_{1 \leq i \leq N} g_i(t), \quad g^*(t) = \max_{1 \leq i \leq N} g_i(t).$$

Theorem 2.4.1

Let the following conditions hold:
   (1)   The function $f(t, y_1, \ldots, y_N)$ is strictly superlinear.
   (2)   Conditions H2.4.1–H2.4.5 hold.

(3) *For any constant $a \neq 0$ the condition*

$$\int_{A(g_*)} |f(t, a, \ldots, a)| \, dt = \infty \tag{2.4.5}$$

holds.

  Then all solutions of equation (2.4.1) oscillate.

Proof

Suppose that $x(t)$ is a nonoscillating solution of equation (2.4.1). Without loss of generality we can assume that $x(t)$ is an eventually positive solution. Introduce the notation

$$y(t) = x(t) + h(t)x(\tau(t)). \tag{2.4.6}$$

Then the function $y(t)$ is also eventually positive and from equation (2.4.1) we obtain that $y(t)$ is a nondecreasing function for t large enough. From (2.4.6) and condition H2.4.5 there follows the inequality

$$
\begin{aligned}
x(t) &= y(t) - h(t)x(\tau(t)) \\
&= y(t) - h(t)[y(\tau(t)) - h(\tau(t)) x(\tau^2(t))] \\
&\geq y(t) - h(t)y(\tau(t)) \geq (1 - \lambda)y(t)
\end{aligned}
\tag{2.4.7}
$$

for t large enough. Then from equation (2.4.1) and (2.4.7) we obtain

$$y'(t) \geq f(t, (1 - \lambda)y(g_1(t)), \ldots, (1 - \lambda)y(g_N(t))), \quad t \geq T,$$

provided $T > t_0$ is large enough. The function $z(t) = (1 - \lambda)y(t)$ then satisfies

$$z'(t) \geq (1 - \lambda)f(t, z(g_1(t)), \ldots, z(g_N(t))), \quad t \geq T. \tag{2.4.8}$$

Since (2.4.5) implies that

$$\int_{t_0}^{\infty} |f(t, a, \ldots, a)| \, dt = \infty \quad \text{for every} \quad a \neq 0 \tag{2.4.9}$$

integrating (2.4.8) and noting that $z(g_i(t))$, $1 \leq i \leq N$, are bounded from below by a positive constant on $[T, \infty)$, we see that $\lim_{t \to \infty} z(t) = \infty$. Hence there exists $T_1 > T$ such that $z(g_*(t)) \geq M$ for $t \geq T_1$. By the strong superlinearity of f we have

$$f(t, z(g_1(t)), \ldots, z(g_N(t))) \geq f(t, z(g_*(t)), \ldots, z(g_*(t)))$$

$$\geq M^{-\alpha}[z(g_*(t))]^\alpha f(t, M, \ldots, M), \quad t \geq T_1$$

which, combined with (2.4.8), gives

$$z'(t) \geq (1 - \lambda)M^{-\alpha}[z(g_*(t))]^\alpha f(t, M, \ldots, M), \quad t \geq T_1 \qquad (2.4.10)$$

Divide (2.4.10) by $[z(t)]^\alpha$ and integrate it over $A(g_*) \cap [T_1, t]$. Since $z(g_*(t)) \geq z(t)$ on this set, we find

$$\int_{T_1}^{t} \frac{z'(s)}{[z(s)]^\alpha} \, ds \geq (1 - \lambda)M^{-\alpha} \int_{A(g_*) \cap [T_1, t]} f(s, M, \ldots, M) \, ds$$

from which, letting $t \to \infty$, we have

$$\int_{A(g_*) \cap [T_1, \infty)} f(s, M, \ldots, M) \, ds \leq \frac{M^\alpha [z(T_1)]^{1-\alpha}}{(1 - \lambda)(\alpha - 1)} < \infty.$$

This contradicts (2.4.5). Theorem 2.4.1 is proved.

Theorem 2.4.2

Let the following conditions hold:
   (1)   The function $f(t, y_1, \ldots, y_N)$ is strictly superlinear.
   (2)   Conditions H2.4.1–H2.4.4 and H2.4.6 hold.
   (3)   For any constant $a \neq 0$ the following condition holds

$$\int_{A(g_*, \tau)} |f(t, a, \ldots, a)| \, dt = \infty. \qquad (2.4.11)$$

Then all solutions of equation (2.4.1) oscillate.

Proof

Suppose that $x(t)$ is a nonoscillating solution of equation (2.4.1). Without loss of generality we can assume that $x(t)$ is an eventually positive solution. Then the function $y(t)$ defined by (2.4.6) is also eventually positive and from equation (2.4.1) we obtain that $y(t)$ is a nondecreasing function for t large enough. From (2.4.6) and condition H2.4.6 there follows the inequality

$$x(t) = \frac{y(\tau^{-1}(t)) - x(\tau^{-1}(t))}{h(\tau^{-1}(t))}$$

$$= \frac{y(\tau^{-1}(t))}{h(\tau^{-1}(t))} - \frac{1}{h(\tau^{-1}(t))}\left[\frac{y(\tau^{-2}(t))}{h(\tau^{-2}(t))} - \frac{x(\tau^{-2}(t))}{h(\tau^{-2}(t))}\right]$$

$$\geq \frac{y(\tau^{-1}(t))}{h(\tau^{-1}(t))} - \frac{y(\tau^{-2}(t))}{h(\tau^{-1}(t))h(\tau^{-2}(t))} \geq \frac{\mu - 1}{\mu v}y(\tau^{-1}(t)) \quad (2.4.12)$$

for t large enough. Then from equations (2.4.1) and (2.4.12) we get

$$y'(t) \geq f\left(t, \frac{\mu - 1}{\mu v}y(\tau_0^{-1}g_1(t)), \ldots, \frac{\mu - 1}{\mu v}y(\tau_0^{-1}g_N(t))\right), \quad t \geq T$$

$T > t_0$ being large enough, which reduces to

$$z'(t) \geq \frac{\mu - 1}{\mu v}f(t, z(\tau_0^{-1}g_1(t)), \ldots, z(\tau_0^{-1}g_N(t))), \quad t \geq T \quad (2.4.13)$$

for $z(t) = (\mu - 1)(\mu v)^{-1}y(t)$. Condition (2.4.11) clearly implies (2.4.9), and so $z(t) \to \infty$ as $t \to \infty$. Choosing $T_1 > t_0$ so that $z(\tau_0^{-1}g_*(t)) \geq M$ for $t \geq T_1$ and proceeding exactly as in Theorem 2.4.1, we obtain

$$\int_{A(g_*, \tau) \cap [T_1, \infty)} f(s, M, \ldots, M)\, ds \leq \frac{\mu v M^\alpha [z(T_1)]^{1-\alpha}}{(\mu - 1)(\alpha - 1)} < \infty$$

which contradicts (2.4.11). This completes the proof of Theorem 2.4.2.

Theorem 2.4.3

Let the following conditions hold:
  (1) *The function $f(t, y_1, \ldots, y_N)$ is strictly sublinear.*
  (2) *Conditions H2.4.1–H2.4.4 and H2.4.7 hold.*
  (3) *For any constant $a \neq 0$ the following condition holds*

$$\int_{R(g^*)} |f(t, a, \ldots, a)|\, dt = \infty. \quad (2.4.14)$$

Then all solutions of equation (2.4.2) oscillate.

Proof

Suppose that $x(t)$ is a nonoscillating solution of equation (2.4.2). Without loss of generality we can assume that $x(t)$ is an eventually positive solution. Then the function $y(t)$ defined by (2.4.6) is also eventually positive and from equation (2.4.2) we obtain that $y(t)$ is a nonincreasing function for t large enough. From (2.4.6) and condition H2.4.7 there follows the inequality (2.4.7). Then from equations (2.4.2) and (2.4.7) we obtain

$$y'(t) + f(t, (1 - \lambda)y(g_1(t)), \ldots, (1 - \lambda)y(g_N(t))) \leq 0, \quad t \geq T$$

which is rewritten for $z(t) = (1 - \lambda)y(t)$ as

$$z'(t) + (1 - \lambda)f(t, z(g_1(t)), \ldots, z(g_N(t))) \leq 0, \quad t \geq T. \quad (2.4.15)$$

From (2.4.15) and (2.4.9) (which is also implied by (2.4.14)) we easily conclude that $\lim_{t \to \infty} z(t) = 0$, and so there exists $T_1 > T$ such that $z(g^*(t)) \leq m$ for $t \geq T_1$. Combining (2.4.15) with the inequality

$$f(t, z(g_1(t)), \ldots, z(g_N(t))) \geq f(t, z(g^*(t)), \ldots, z(g^*(t)))$$

$$\geq m^{-\beta}[z(g^*(t))]^\beta f(t, m, \ldots, m), \quad t \geq T$$

we obtain

$$-z'(t) \geq (1 - \lambda)m^{-\beta}[z(g^*(t))]^\beta f(t, m, \ldots, m), \quad t \geq T_1. \quad (2.4.16)$$

Divide (2.4.16) by $[z(t)]^\beta$, integrate it over $R(g^*) \cap [T_1, t]$ and then let $t \to \infty$. Noting that $z(g^*(t)) \geq z(t)$ on $R(g^*) \cap [T_1, \infty)$, we then have

$$\int_{R(g^*) \cap [T_1, \infty)} f(s, m, \ldots, m) \, ds \leq \frac{m^\beta [z(T_1)]^{1-\beta}}{(1 - \lambda)(1 - \beta)} < \infty$$

which contradicts (2.4.14). This completes the proof of Theorem 2.4.3.

Theorem 2.4.4

Let the following conditions hold:
  (1) The function $f(t, y_1, \ldots, y_N)$ is strictly sublinear.
  (2) Conditions H2.4.1–H2.4.4 and H2.4.8 hold.

(3)   *For any constant $a \neq 0$ the following condition holds*

$$\int\limits_{R(g^*,\, \tau)} |f(t, a, \ldots, a)| \, dt = \infty \qquad (2.4.17)$$

Then all solutions of equation (2.4.2) oscillate.

Proof

Suppose that $x(t)$ is a nonoscillating solution of equation (2.4.2). Without loss of generality we can assume that $x(t)$ is an eventually positive solution. Then the function $y(t)$ defined by (2.4.6) is also eventually positive and from equation (2.4.2) we obtain that $y(t)$ is a nonincreasing function for t large enough. From (2.4.6) and condition H2.4.8 there follows inequality (2.4.12). Then from equation (2.4.2) and (2.4.12) we obtain that $z(t) = (\mu - 1)y(t)(\mu v)^{-1}$ satisfies the inequality

$$z'(t) + \frac{\mu - 1}{\mu v} f(t, z(\tau_0^{-1} g_1(t)), \ldots, z(\tau_0^{-1} g_N(t))) \leqq 0, \quad t \geqq T \qquad (2.4.18)$$

provided $T \geqq t_0$ is large enough. Noting that $\lim_{t \to \infty} z(t) = 0$ because of (2.4.17) and (2.4.18) and applying the same argument as above, we are led to the inequality

$$\int\limits_{R(g^*,\, \tau) \cap [T_1,\, \infty)} f(s, m, \ldots, m) \, ds \leqq \frac{\mu v m^\beta [z(T_1)]^{1-\beta}}{(\mu - 1)(1 - \beta)} < \infty$$

for some $T_1 > T$, which contradicts (2.4.17). This completes the proof of Theorem 2.4.4.

In the subsequent theorems necessary and sufficient conditions for existence of bounded nonoscillating solutions of equations (2.4.1) and (2.4.2) are obtained. Denote by N the set of all nonoscillating solutions of equations (2.4.1) or (2.4.2). Introduce the following notation:

$$N_\infty = \left\{ x \in N : \lim_{t \to \infty} |x(t) + h(t)x(\tau(t))| = \infty \right\},$$

$$N_c = \left\{ x \in N : \lim_{t \to \infty} |x(t) + h(t)x(\tau(t))| = \text{const.} > 0 \right\},$$

$$N_0 = \left\{ x \in N : \lim_{t \to \infty} |x(t) + h(t)x(\tau(t))| = 0 \right\}.$$

From (2.4.1) and (2.4.2) it follows immediately that
$$N = N_c \cup N_\infty \text{ for equation (2.4.1)}$$
and
$$N = N_0 \cup N_c \text{ for equation (2.4.2)}.$$

Theorem 2.4.5

Let conditions (H2.4) hold in addition to condition H2.4.5 or H2.4.6. Then a necessary and sufficient condition for equation (2.4.1) to have a bounded nonoscillating solution is the condition

$$\int_{t_0}^{\infty} |f(t, b, \ldots, b)| \, dt < \infty \tag{2.4.19}$$

for some constant $b \neq 0$.

Proof

Necessity. Let $x(t)$ be a bounded nonoscillating solution of (2.4.1). We can assume that $x(t)$ is eventually positive. Put $y(t) = x(t) + h(t)x(\tau(t))$. Then (2.4.7) or (2.4.12) holds depending on whether $x(t)$ satisfies H2.4.5 or H2.4.6. In either case, $y(t)$ is bounded and nondecreasing, so that there exist constants $c > 0$ and $T > t_0$ such that

$$x(g_i(t)) \geq c \quad \text{for } t \geq T, \quad 1 \leq i \leq N. \tag{2.4.20}$$

An integration of (2.4.1) then shows that

$$\int_{T}^{\infty} f(t, c, \ldots, c) \, dt \leq y(T) - y(\infty) < y(T) < \infty,$$

which implies (2.4.19).

Sufficiency. The proof of this part is an extended adaptation of the method introduced by Ruan [138]. We can assume without loss of generality that the constant b in (2.4.19) is positive.

(i) Let condition H2.4.5 hold. Fix a positive constant $c \leq b$ and choose $T > t_0$ large enough so that

$$T_0 = \min\left\{\tau(T), \inf_{t \geq T} g_1(t), \ldots, \inf_{t \geq T} g_N(t)\right\} \geq t_0 \tag{2.4.21}$$

and

$$\int_T^\infty f(t, c, \ldots, c) \, dt \le (1 - \lambda)c. \tag{2.4.22}$$

Denote by X the set

$$X = \{x \in C[T_0, \infty): \lambda c \le x(t) \le c \quad \text{for } t \ge T,$$
$$x(t) = x(T) \quad \text{for } T_0 \le t \le T\}. \tag{2.4.23}$$

With each $x \in X$ we associate the function $\hat{x}: [T_0, \infty) \to \mathbb{R}$ defined by the formula

$$\begin{cases} \hat{x}(t) = \sum_{i=0}^{n(t)-1} (-1)^i H_i(t) x(\tau^i(t)) + \dfrac{(-1)^{n(t)} x(T)}{1 + h(T)} H_{n(t)}(t), & t > T, \\[2mm] \hat{x}(t) = \dfrac{x(T)}{1 + h(T)}, & T_0 \le t \le T, \end{cases}$$
$$\tag{2.4.24}$$

where $n(t)$ is the least positive integer such that $T_0 < \tau^{n(t)}(t) \le T$ and

$$H_0(t) \equiv 1, \quad H_i(t) = \prod_{j=0}^{i-1} h(\tau^j(t)), \quad i = 1, 2, \ldots. \tag{2.4.25}$$

It can be shown that $\hat{x}(t)$ is continuous and positive for $t \ge T_0$ and satisfies the functional equation

$$\hat{x}(t) + h(t)\hat{x}(\tau(t)) = x(t), \quad t \ge T \tag{2.4.26}$$

In fact, the continuity of $\hat{x}(t)$ and equation (2.4.26) are easily verified, and the positivity of $\hat{x}(t)$ follows from (2.4.24) rewritten as

$$\hat{x}(t) = \sum_{j=0}^{m-1} [(-1)^{2j} H_{2j}(t) x(\tau^{2j}(t)) + (-1)^{2j+1} H_{2j+1}(t) x(\tau^{2j+1}(t))]$$

$$+ (-1)^{2m} H_{2m}(t) x(\tau^{2m}(t)) + \frac{(-1)^{2m+1} x(T)}{1 + h(T)} H_{2m+1}(t)$$

$$\text{for } n(t) = 2m + 1, \quad m = 0, 1, \ldots,$$

$$\hat{x}(t) = \sum_{j=0}^{m-1} [(-1)^{2j} H_{2j}(t) x(\tau^{2j}(t)) + (-1)^{2j+1} H_{2j+1}(t) x(\tau^{2j+1}(t))]$$

$$+ \frac{(-1)^{2m} x(T)}{1 + h(T)} H_{2m}(T) \quad \text{for } n(t) = 2m, \quad m = 1, 2, \ldots,$$

combined with the observation that, in view of (2.4.23), $x \in X$ implies that

$$(-1)^{2j}H_{2j}(t)x(\tau^{2j}(t)) + (-1)^{2j+1}H_{2j+1}(t)x(\tau^{2j+1}(t))$$
$$= H_{2j}(t)[x(\tau^{2j}(t)) - h(\tau^{2j}(t))x(\tau^{2j+1}(t))]$$
$$\geq H_{2j}(t)[\lambda c - \lambda c] = 0, \quad t \geq T$$

and

$$(-1)^{2m}H_{2m}(t)x(\tau^{2m}(t)) + \frac{(-1)^{2m+1}x(T)}{1+h(T)}H_{2m+1}(t)$$
$$\geq H_{2m}(t)[x(\tau^{2m}(t)) - h(\tau^{2m}(t))x(T)]$$
$$\geq H_{2m}(t)[\lambda c - \lambda c] = 0, \quad t \geq T.$$

Now define the mapping $F : X \to C[T_0, \infty)$ by

$$\begin{cases} Fx(t) = c - \displaystyle\int_t^\infty f(s, \hat{x}(g_1(s)), \ldots, \hat{x}(g_N(s))) \, ds, \quad t \geq T \\[4mm] Fx(t) = c - \displaystyle\int_T^\infty f(s, \hat{x}(g_1(s)), \ldots, \hat{x}(g_N(s))) \, ds, \quad T_0 \leq t \leq T \end{cases} \qquad (2.4.27)$$

Noting that $x \in X$ implies $0 \leq \hat{x}(t) \leq x(t) \leq c$, $t \geq T_0$, because of (2.4.26), we see from (2.4.27) and (2.4.22) that

$$c \geq Fx(t) \geq c - \int_T^\infty f(s, c, \ldots, c) \, ds \geq c - (1-\lambda)c = \lambda c, \quad t \geq T_0,$$

for every $x \in X$. Thus F maps X into itself. It can be proved in a standard way that F is continuous and $F(X)$ is relatively compact in the topology of the Fréchet space $C[T_0, \infty)$. Therefore, by the Schauder–Tychonoff fixed point theorem, there exists $x \in X$ such that $x = Fx$, that is

$$x(t) = c - \int_t^\infty f(s, \hat{x}(g_1(s)), \ldots, \hat{x}(g_N(s))) \, ds, \quad t \geq T$$

which, because of (2.4.26), can be rewritten as

$$\dot{x}(t) + h(t)\dot{x}(\tau(t)) = c - \int_t^\infty f(s, \hat{x}(g_1(s)), \dots, \hat{x}(g_N(s))) \, ds, \quad t \geq T.$$

$$(2.4.28)$$

Differentiation of (2.4.28) shows that the function $\hat{x}(t)$ associated with $x(t)$ by (2.4.24) satisfies equation (2.4.1) for $t \geq T$. Since $\hat{x}(t)$ satisfies

$$(1 - \lambda)x(t) \leq \hat{x}(t) \leq x(t) \qquad (2.4.29)$$

for all large t, $\hat{x}(t)$ is a bounded nonoscillating solution of (2.4.1).

(ii) Let condition H2.4.6 hold. Let $c > 0$ be such that $\mu c \leq b$ and choose $T > t_0$ so that (2.4.21) with $\tau(T)$ replaced by $\tau^{-1}(T)$ and the following inequality hold:

$$\int_T^\infty f(t, \mu c, \dots, \mu c) \, dt \leq (\mu - 1)c. \qquad (2.4.30)$$

The set $X \subset C[T_0, \infty)$ and the mapping $F : X \to C[T_0, \infty)$ are defined as follows:

$$X = \{x \in C[T_0, \infty) : c \leq x(t) \leq \mu c \text{ for } t \geq T, \quad x(t) = x(T) \text{ for } T_0 \leq t \leq T\}$$

$$(2.4.31)$$

and

$$\begin{cases} Fx(t) = \mu c - \displaystyle\int_t^\infty f(s, \hat{x}(g_1(s)), \dots, \hat{x}(g_N(s))) \, ds, \quad t \geq T \\[4mm] Fx(t) = \mu c - \displaystyle\int_T^\infty f(s, \hat{x}(g_1(s)), \dots, \hat{x}(g_N(s))) \, ds, \quad T_0 \leq t \leq T \end{cases}$$

$$(2.4.32)$$

where $\hat{x}(t)$ denotes the function constructed from $x \in X$ by the rule

$$\hat{x}(t) = \sum_{i=1}^{n(t)} \frac{(-1)^{i-1}x(\tau^{-i}(t))}{H_i(\tau^{-i}(t))} + \frac{(-1)^{n(t)}x(T)}{(1 + h(T))H_{n(t)}(\tau^{-n(t)}(t))}, \quad t > T$$

$$(2.4.33)$$

$$\hat{x}(t) = \frac{x(T)}{1 + h(T)}, \quad T_0 \leq t \leq T,$$

$n(t)$ being the least positive integer such that $T_0 < \tau^{-n(t)}(t) \leq T$. We note that, for every $x \in X$, $\hat{x}(t)$ is continuous and positive for $t \geq T_0$ and satisfies (2.4.26) for $t \geq T$. Then, proceeding as in the above case, we can show that F has a fixed element $x \in X$ which gives rise to a solution $\hat{x}(t)$ of (2.4.1) on $[T, \infty)$. That $\hat{x}(t)$ is bounded and nonoscillating follows from the relation

$$\frac{\mu - 1}{\mu \nu} x(\tau^{-1}(t)) \leq \hat{x}(t) \leq x(t), \quad t \geq T, \tag{2.4.34}$$

(see (2.4.12)). This completes the proof of Theorem 2.4.5.

Theorem 2.4.6

Let conditions (H2.4) hold in addition to condition H2.4.7 or H2.4.8. Then a necessary and sufficient condition for equation (2.4.2) to have a bounded nonoscillating solution is condition (2.4.19).

Proof

The necessity follows readily from (2.4.2) integrated over a neighbourhood of infinity.

To prove the sufficiency we proceed as follows.

(i) Let condition H2.4.7 hold. Choose $c > 0$ and $T > t_0$ so that $c \leq b$ and (2.4.21) (with $\tau(T)$ replaced by $\tau^{-1}(T)$) and (2.4.22) hold, let $X \subset C[T_0, \infty)$ be as in (2.4.23) and define $F: X \to C[T_0, \infty)$ by

$$\begin{cases} Fx(t) = \lambda c + \displaystyle\int_t^\infty f(s, \hat{x}(g_1(s)), \ldots, \hat{x}(g_N(s)))\, ds, & t \geq T \\[4mm] Fx(t) = \lambda c + \displaystyle\int_T^\infty f(s, \hat{x}(g_1(s)), \ldots, \hat{x}(g_N(s)))\, ds, & T_0 \leq t \leq T \end{cases} \tag{2.4.35}$$

where $\hat{x}: [T_0, \infty) \to \mathbb{R}$ is given by

$$\begin{cases} \hat{x}(t) = \displaystyle\sum_{i=0}^\infty (-1)^i H_i(t) x(\tau^i(t)), & t \geq T, \\[4mm] \hat{x}(t) = \hat{x}(T), & T_0 \leq t \leq T. \end{cases} \tag{2.4.36}$$

First we check that $x \in X$ implies that $\hat{x}(t)$ is continuous and positive on $[T_0, \infty)$ and satisfies (2.4.26) on $[T, \infty)$, and then we apply the Schauder–

Tychonoff theorem to establish the existence of a fixed element $x \in X$ of F. The associated function $\hat{x}(t)$ is shown to satisfy equation (2.4.2) on $[T, \infty)$. The boundedness of $x(t)$ follows from (2.4.29) holding also in this case.

(ii) Let condition H2.4.8 hold. Let $c > 0$ and $T > t_0$ be such that $\mu c \leqq b$ and (2.4.21) and (2.4.30) hold. Consider the set X defined by (2.4.31) and associate with every $x \in X$ the function $\hat{x}(t)$:

$$\begin{cases} \hat{x}(t) = \sum_{i=1}^{\infty} \dfrac{(-1)^{i-1} x(\tau^{-i}(t))}{H_i(\tau^{-i}(t))}, & t \geqq T, \\ \hat{x}(t) = \hat{x}(T), & T_0 \leqq t \leqq T. \end{cases} \qquad (2.4.37)$$

The desired solution of (2.4.2) will be obtained as the fuction $\hat{x}(t)$ associated with a fixed point $x \in X$ of the mapping F defined by

$$\begin{cases} Fx(t) = c + \displaystyle\int_t^{\infty} f(s, \hat{x}(g_1(s)), \ldots, \hat{x}(g_N(s))) \, ds, & t \geqq T, \\ Fx(t) = c + \displaystyle\int_T^{\infty} f(s, \hat{x}(g_1(s)), \ldots, \hat{x}(g_N(s))) \, ds, & T_0 \leqq t \leqq T. \end{cases} \qquad (2.4.38)$$

Note that (2.4.34) holds, which ensures the boundedness of the solution $x(t)$. The proof of Theorem 2.4.6 is thus complete.

A corollary of Theorems 2.4.1, 2.4.2 and 2.4.5 is the following necessary and sufficient condition for oscillation of the solutions of equation (2.4.1).

Theorem 2.4.7

Let the following conditions hold:
(1) The function $f(t, y_1, \ldots, y_N)$ is strictly superlinear.
(2) Conditions (H2.4) hold.
(3) Condition H2.4.5 (H2.4.6) holds in addition to

$$g_*(t) > t \quad (g_*(t) > \tau(t)) \text{ for } t \text{ large enough.}$$

Then a necessary and sufficient condition for all solutions of equation (2.4.1) to oscillate is the condition

$$\int_{t_0}^{\infty} |f(t, a, \ldots, a)| \, dt = \infty \qquad (2.4.39)$$

for every constant $a \neq 0$.

A corollary of Theorems 2.4.3, 2.4.4 and 2.4.6 is the following necessary and sufficient condition for oscillation of the solutions of equation (2.4.2).

Theorem 2.4.8

Let the following conditions hold:
  (1)   *The function $f(t, y_1, \ldots, y_N)$ is strictly sublinear.*
  (2)   *Conditions ($H2.4$) hold.*
  (3)   *Condition $H2.4.7$ ($H2.4.8$) holds in addition to*

$$g^*(t) < t \quad (g^*(t) < \tau(t)) \text{ for } t \text{ large enough.}$$

Then a necessary and sufficient condition for all solutions of equation (2.4.2) to oscillate is condition (2.4.39).

Example 2.4.1   Using the notation $F_\gamma(u) = |u|^\gamma \operatorname{sgn} u$, consider the equation

$$\frac{d}{dt}[x(t) + \lambda x(t - \rho)] = p(t)F_\gamma(x(t + \sigma)) + q(t)F_\delta(x(\theta t)) \qquad (2.4.40)$$

where γ, δ, λ, ρ, σ and θ are positive constants and $p, q : [t_0, \infty) \to [0, \infty)$ are continuous functions. We assume that $\lambda < 1$ and neither $p(t)$ nor $q(t)$ vanishes identically. Then, (2.4.40) is a special case of (2.4.1) in which $N = 2$, $h(t) \equiv \lambda$, $\tau(t) = t - \rho$, $g_1(t) = t + \sigma$, $g_2(t) = \theta t$ and

$$f(t, y_1, y_2) = p(t)F_\gamma(y_1) + q(t)F_\delta(y_2). \qquad (2.4.41)$$

The function (2.4.41) is strongly superlinear if $\gamma > 1$ and $\delta > 1$ (the superlinearity constant is taken to be $\alpha = \min\{\gamma, \delta\}$).
  (i) By Theorem 2.4.5, equation (2.4.40) has a bounded nonoscillating solution if and only if

$$\int_{t_0}^\infty p(t)\,dt < \infty \quad \text{and} \quad \int_{t_0}^\infty q(t)\,dt < \infty. \qquad (2.4.42)$$

  (ii) Suppose that $\gamma > 1$, $\delta > 1$ and $\theta > 1$. Then from Theorem 2.4.7 it follows that all solutions of (2.4.40) are oscillatory if and only if

$$\int_{t_0}^\infty p(t)\,dt = \int_{t_0}^\infty q(t)\,dt = \infty. \qquad (2.4.43)$$

Remark 2.4.1 It remains to characterize the class N_∞ of unbounded nonoscillating solutions of (2.4.1) and the class N_0 of decaying nonoscillating solutions of (2.4.2). This, however, seems to be a difficult problem since N_∞ and N_c (N_0 and N_c) may or may not coexist as the following example shows.

Consider the equation

$$\frac{d}{dt}[x(t) + \lambda x(t - \rho)] = e^{\gamma \sigma}(1 + \lambda e^{-\rho}) e^{(1 - \gamma)t} F_\gamma(x(t - \sigma)), \qquad (2.4.44)$$

where $\lambda > 0$, $\gamma > 0$, ρ and σ are constants. Equation (2.4.44) is a special case of (2.4.1) if $0 < \lambda < 1$ and $\rho > 0$ (respectively $\lambda > 1$ and $\rho < 0$). In either of such cases, Theorem 2.4.5 shows that $N_c \neq \varnothing$ for (2.4.44) if $\gamma > 1$ and $N_c = \varnothing$ for (2.4.44) if $\gamma \leq 1$. On the other hand, one easily sees that (2.4.44) has an unbounded solution $x(t) = e^t$ belonging to the class N_∞ for any values of γ, λ, ρ and σ.

2.5 Oscillation and comparison results in neutral differential equations and their applications to the delay logistic equation

A wide class of population growth models are described by the logistic equation

$$x'(t) = rx(t)\left[1 - \frac{x(t)}{K}\right] \qquad (2.5.1)$$

where the intrinsic growth rate r and the carrying capacity K are positive real numbers. The general solution of equation (2.5.1) can be given in the form

$$x(t) = \frac{K}{1 + \left(\dfrac{K}{x_0} - 1\right)\exp(-rt)}, \qquad x_0 > 0 \qquad (2.5.2)$$

which shows that every solution of equation (2.5.1) is a monotone function and tends to the steady state solution K of equation (2.5.1), as $t \to +\infty$.

However, when examining the growth curves of some populations, one sees that there are fluctuations about the steady state K. This is attributed to a number of biological phenomena; nonhomogeneity of the population, variable growth rates of the members, or the population has a 'memory' which means that the per capita growth rate is associated with some earlier state of the population. In this section we concentrate on the last case, that is when the model equation is the delay logistic equation

$$x'(t) = rx(t)\left[1 - \frac{\sum_{i=1}^{n} p_i x(t - \tau_i)}{K}\right] \qquad (2.5.3)$$

where the weight constants p_i and the delays τ_i are positive constants, $\sum_{i=1}^{n} p_i = 1$. For a more complete treatment, see Nisbet and Gurney [126].

When the density-dependent regulation of vital rate $x'(t)$ operates with some constant delay, Györi and Witten [59] introduced the so-called delay logistic equation with growth rate lags:

$$x'(t) = \sum_{j=1}^{m} q_j x'(t - \sigma_j) + rx(t)\left[1 - \frac{\sum_{i=1}^{n} p_i x(t - \tau_i)}{K}\right] \qquad (2.5.4)$$

where the weight constants q_j and the delays σ_j are positive. Equations (2.5.2) and (2.5.3) have the same positive steady state K. A natural question arises now, namely, do the solutions of those equations oscillate around the steady state K?

The purpose of this section is to answer that question. In the case of equation (2.5.3), we prove that all solutions of equation (2.5.3) oscillate around K if and only if the linear equation

$$x'(t) = -r \sum_{i=1}^{n} p_i x(t - \tau_i) \qquad (2.5.5)$$

is oscillatory. This improves the earlier results of Gopalsamy [37], [38] and Kulenovic *et al* [94]. For the neutral equation (2.5.4) we prove that it has a solution $x(t)$ such that $x(t) - K > 0$ provided that the equation

$$x'(t) = \sum_{j=1}^{m} q_j x'(t - \sigma_j) - r \sum_{i=1}^{n} p_i x(t - \tau_i) \qquad (2.5.6)$$

has a positive solution. Conversely, we can prove that if equation (2.5.4) has a solution $x(t)$ such that $x(t) - K > 0$ and $x'(t) < 0$, then equation (2.5.6) has a positive solution.

The second part of the section contains some new comparison results for neutral equations and we give the necessary and sufficient conditions for the linear neutral inequality in addition to those for the equation with only oscillating solutions. For instance, from these results we have that if $q_j > 0$, $\sigma_j > 0$ $(1 \le j \le m)$, $r > 0$, $p_i > 0$, $\tau_i > 0$ $(1 \le i \le n)$ and $\sum_{j=1}^{m} q_j < 1$, then equation (2.5.6) has only an oscillating solution if and only if the equation

$$\lambda = \lambda \sum_{j=1}^{m} q_j \exp(\lambda\sigma_j) + r \sum_{i=1}^{n} p_i \exp(\lambda\tau_i) \qquad (2.5.7)$$

has no real root.

The advantage of this result is that for equation (2.5.7) many conditions are known to have no real root as well as to have a real root (see, for example [6], [33], [55], [94], [102], [105], [161] and the references cited therein).

In the subsequent theorems some earlier results of Myshkis [123], Driver [28] and Grove *et al* [55] are generalized.

Consider the delay differential inequalities of neutral type of the form

$$x'(t) < \sum_{j=1}^{m} q_j(t)x'(t - \sigma_j(t)) - \sum_{i=1}^{n} p_i(t)x(t - \tau_i(t)) \qquad (2.5.8)$$

and

$$y'(t) \geq \sum_{j=1}^{m} q_j(t)y'(t - \sigma_j(t)) - \sum_{i=1}^{n} p_i(t)y(t - \tau_i(t)) \qquad (2.5.9)$$

on the interval $[t_0, T]$, where $-\infty < t_0 < T \leq \infty$. Assume the following conditions fulfilled:

H2.5.1 $p_i \in C([t_0, \infty); \mathbb{R}_+)$ and $\tau_i \in C([t_0, \infty); \mathbb{R}_+)$, $(1 \leq i \leq n)$,

moreover,

$$\tau = \sup_{t \geq t_0} \left\{ \max_{1 \leq i \leq n} \tau_i(t) \right\} < \infty;$$

H2.5.2 $q_j \in BM([t_0, \infty); \mathbb{R}_+)$ and $\sigma_j \in C([t_0, \infty); \mathbb{R}_+)$, $(1 \leq j \leq m)$,

$$\sigma = \sup_{t \geq t_0} \left\{ \max_{1 \leq j \leq m} \sigma_j(t) \right\} < \infty$$

where $BM([t_0, \infty); \mathbb{R}_+)$ is the set of the locally bounded and Lebesgue measurable functions $u: [t_0, \infty) \to \mathbb{R}_+$.

Definition 2.5.1 The functions $x: [t_0 - \gamma, T) \to \mathbb{R}$ and $y: [t_0 - \gamma, T) \to \mathbb{R}$, $(\gamma = \max\{\tau, \sigma\})$, are said to be absolutely continuous (continuously differentiable) solutions of equations (2.5.8) and (2.5.9), respectively, if $x(t)$ and $y(t)$ are continuous on $[t_0 - \gamma, T)$, their derivatives $x'(t)$ and $y'(t)$ exist almost everywhere (a.e.) on $[t_0 - \sigma, T)$ and $x', y' \in BM([t_0 - \sigma, T); \mathbb{R})$ ($x(t)$ and $y(t)$ are continuously differentiable on $[t_0 - \sigma, T)$). Moreover, $x(t)$ and $y(t)$ satisfy equations (2.5.8) and (2.5.9), respectively, a.e. on $[t_0, T)$.

$x(t)$ and $y(t)$ are said to be positive solutions if $x(t) > 0$ and $y(t) > 0$ on $t_0 - \gamma \leq t < T$.

Theorem 2.5.1

Assume that H2.5.1 and H2.5.2 hold, moreover,

$$\sum_{i=1}^{n} p_i(t)\tau_i(t) > 0 \quad \text{and} \quad \sigma_j(t) > 0, \quad (1 \leq j \leq n, t_0 \leq t \leq T). \quad (2.5.10)$$

Suppose further that $x(t)$ is an absolutely continuous and positive solution of equation (2.5.8) on $[t_0 - \gamma, T)$ such that

$$x'(t) \leq 0, \quad \text{a.e. on } [t_0 - \sigma, T). \quad (2.5.11)$$

If $y(t)$ is an absolutely continuous solution of equation (2.5.9) on $[t_0 - \gamma, T)$ such that

$$y(t) > 0, \quad t_0 - \sigma \leq t \leq t_0 \quad (2.5.12)$$

$$\frac{x'(t)}{x(t)} \leq \frac{y'(t)}{y(t)}, \quad \text{a.e. on } [t_0 - \sigma, t_0] \quad (2.5.13)$$

and

$$\frac{y(t)}{x(t)} \leq \frac{y(t_0)}{x(t_0)}, \quad t_0 - \gamma \leq t \leq t_0, \quad (2.5.14)$$

then

$$\frac{y(t)}{x(t)} > \frac{y(t_0)}{x(t_0)} \quad (2.5.15)$$

on $[t_0, T)$.
We shall use the following lemma.

Lemma 2.5.1

Assume that H2.5.1 and H2.5.2 hold. Suppose further that $x(t)$ and $y(t)$ are absolutely continuous solutions of equations (2.5.8) and (2.5.9), respectively, on $[t_0 - \gamma, T)$ *such that* $x(t) > 0$ *and* $y(t) > 0$ *on* $[t_0 - \sigma, T_1)$. *where* $T_1 \in (t_0, T)$.

Then

$$\alpha(t) = -\frac{x'(t)}{x(t)} \quad \text{and} \quad \beta(t) = -\frac{y'(t)}{y(t)}, \quad t_0 - \sigma \leq t < T_1 \quad (2.5.16)$$

satisfy the following inequalities:

$$\alpha(t) > \sum_{j=1}^{m} q_j(t)\alpha(t - \sigma_j(t)) \exp\left(\int_{t-\sigma_j(t)}^{t} \alpha(u)\, du \right)$$

$$+ \sum_{i=1}^{n} p_i(t) \frac{x(t - \tau_i(t))}{x(\gamma_i(t))} \exp\left(\int_{\gamma_i(t)}^{t} \alpha(u)\, du \right) \quad (2.5.17)$$

and

$$\beta(t) \leq \sum_{j=1}^{m} q_j(t)\beta(t - \sigma_j(t)) \exp\left(\int_{t-\sigma_j(t)}^{t} \beta(u)\, du \right)$$

$$+ \sum_{i=1}^{n} p_i(t) \frac{y(t - \tau_i(t))}{y(\gamma_i(t))} \exp\left(\int_{\gamma_i(t)}^{t} \beta(u)\, du \right) \quad (2.5.18)$$

respectively, a.e. on $[t_0, T_1)$, *where*

$$\gamma_i(t) = \max\{t_0, t - \tau_i(t)\}, \quad 1 \leq i \leq n, \quad t_0 \leq t < T_1. \quad (2.5.19)$$

Proof

One can see that

$$\frac{x(s)}{x(t)} = \exp\left(-\int_{s}^{t} \frac{x'(u)}{x(u)}\, du \right) = \exp\left(\int_{s}^{t} \alpha(u)\, du \right), \quad t_0 - \sigma \leq s \leq t < T_1,$$

and from equation (2.5.8), it follows that

$$\frac{-x'(t)}{x(t)} > \sum_{j=1}^{m} q_j(t) \frac{-x'(t - \sigma_j(t))}{x(t - \sigma_j(t))} \frac{x(t - \sigma_j(t))}{x(t)}$$

$$+ \sum_{i=1}^{n} p_i(t) \frac{x(t - \tau_i(t))}{x(\gamma_i(t))} \frac{x(\gamma_i(t))}{x(t)},$$

which implies the required inequality (2.5.17). The proof of inequality (2.5.18) proceeds in the same way.

Proof of Theorem 2.5.1

We complete the proof in two steps. First, we show that (2.5.15) holds on any interval $[t_0, T_1)$, $(T_1 \in (t_0, T])$, if $y(t) > 0$, $t_0 \leq t < T_1$. Second, we show that $y(t) > 0$ on $[t_0, T)$.

(a) Assume $y(t) > 0$ on some interval $[t_0, T_1)$, where $t_0 < T_1 \leq T$. The proof of the validity of (2.5.15) on $[t_0, T_1]$ will be complete if we show that $\mu(t) = 0$ on $[t_0, T_1)$, where

$$\mu(t) = m(\{t_0 \leq s \leq t : \alpha(s) \leq \beta(s)\}) \qquad (2.5.20)$$

and $m(X)$ is the Lesbesgue measure of the set X. In fact, in that case

$$\ln \frac{x(t_0)}{x(t)} = -\int_{t_0}^{t} \frac{x'(s)}{x(s)} \, ds = \int_{t_0}^{t} \alpha(s) \, ds > \int_{t_0}^{t} \beta(s) \, ds = -\int_{t_0}^{t} \frac{y'(s)}{y(s)} \, ds = \ln \frac{y(t_0)}{y(t)}$$

that is equation (2.5.15) holds on $[t_0, T_1)$.

Now assume that $\mu(t) = 0$ does not hold for any $t \in [t_0, T_1)$, that is, the set

$$U = \{t \in [t_0, T_1) : \mu(t) > 0\}$$

is nonempty. Let $t_1 = \inf U$. Since $\mu(t)$ is a monotone nondecreasing function, it is clear that

$$\mu(t) = 0, \quad t_0 \leq t < t_1 \qquad \text{and} \qquad \mu(t) > 0, \quad t_1 < t < T. \quad (2.5.21)$$

Since $\tau_i \in C([t_0, T]; \mathbb{R}_+)$ and $\alpha, \beta \in BM([t_0, T_1); \mathbb{R})$, one can see that $\gamma_i \in C([t_0, t); \mathbb{R})$ and the functions

$$\int_{\gamma_i(t)}^{t} \alpha(u) \, du \qquad \text{and} \qquad \int_{\gamma_i(t)}^{t} \beta(u) \, du, \quad 1 \leq i \leq n$$

are continuous on $[t_0, T_1)$.

Thus the functions

$$A(t) \equiv \sum_{i=1}^{m} p_i(t) \frac{x(t - \tau_i(t))}{x(\gamma_i(t))} \exp \left(\int_{\gamma_i(t)}^{t} \alpha(u) \, du \right)$$

and

$$B(t) \equiv \sum_{i=1}^{m} p_i(t) \frac{y(t - \tau_i(t))}{y(\gamma_i(t))} \exp \left(\int_{\gamma_i(t)}^{t} \beta(u) \, du \right)$$

are continuous on $[t_0, T)$.

From the definition of t_1 it follows that

$$\int_{\gamma_i(t_1)}^{t_1} \alpha(u) \, du \geq \int_{\gamma_i(t_1)}^{t_1} \beta(u) \, du, \quad 1 \leq i \leq n.$$

Moreover, by condition (2.5.10) there exists an index t_0 such that $p_{i_0}(t_1)\tau_{i_0}(t_1) > 0$. But in that case $\gamma_{i_0}(t_1) > 0$ and (2.5.21) implies

$$p_{i_0}(t_1) \exp \left(\int_{\gamma_{i_0}(t_1)}^{t_1} \alpha(u) \, du \right) > p_{i_0}(t_1) \exp \left(\int_{\gamma_{i_0}(t_1)}^{t_1} \beta(u) \, du \right).$$

On the other hand, condition (2.5.14) and the definition of γ_i yield

$$\frac{x(t - \tau_i(t))}{x(\gamma_i(t))} \geq \frac{y(t - \tau_i(t))}{y(\gamma_i(t))} > 0, \quad t_0 \leq t < T_1, \quad 1 \leq i \leq n,$$

therefore $A(t_1) > B(t_1)$.

Now, by condition (2.5.10) we have

$$\min\{A(t_1) - B(t_1), \min_{1 \leq i \leq n} \sigma_i(t_1)\} > 0,$$

moreover, there exists $\delta > 0$ such that

$$A(t) > B(t) \quad \text{and} \quad t - \sigma_i(t) < t_1, \quad 1 \leq i \leq n, \qquad (2.5.22)$$

for any $t \in [t_1, t_1 + \delta)$.

Since

$$\alpha(t) > \beta(t) \quad \text{and} \quad \alpha(t) = -\frac{x'(t)}{x(t)} \geq 0, \quad \text{a.e. on } [t_0 - \sigma, t_1)$$

we obtain

$$q_j(t)\alpha(t - \sigma_j(t)) \exp\left(\int_{t - \sigma_j(t)}^{t} \alpha(u) \, du \right)$$

$$\geq q_j(t)\beta(t - \sigma_j(t)) \exp\left(\int_{t - \sigma_j(t)}^{t} \beta(s) \, ds \right), \quad 1 \leq j \leq n$$

for almost every $t \in [t_1, t_1 + \delta)$.

Thus conditions (2.5.15), (2.5.16) and (2.5.22) imply $\alpha(t) > \beta(t)$ for almost every $t \in [t_1, t_1 + \delta)$, that is $\mu(t) = m(\{t_0 \leq s \leq t : \alpha(s) \leq \beta(s)\}) = 0$ on $[t_1, t_1 + \delta)$. But this is incompatible with condition (2.5.21); therefore, the set U is empty and the result follows.

(b) Now, we prove that $y(t) > 0$ on $t_0 \leq t < T$. Assume that $y(t) > 0$ does not hold for any $t \in [t_0, T)$. Then the set

$$V = \{t_0 \leq t \leq T : y(t) = 0\}$$

is nonempty. Let $T_1 = \inf V$. Then $T_1 > t_0$, since $y(t_0) > 0$ and $y(t)$ is continuous.

Moreover,

$$y(t) > 0, \quad t_0 \leq t < T_1 \quad \text{and} \quad y(T_1) = 0.$$

But, in that case, we proved that condition (2.5.15) is valid on $[t_0, T_1)$, that is

$$y(t) > \frac{y(t_0)}{x(t_0)} x(t), \quad t_0 \leq t < T_1.$$

Thus

$$y(T_1) = \lim_{t \to T_1} y(t) \geq \lim_{t \to T_1} \frac{y(t_0)}{x(t_0)} x(t) = \frac{y(t_0)}{x(t_0)} x(T_1) > 0$$

which contradicts condition (2.5.23).

Thus $y(t)$ is positive on $[t_0, T)$ and condition (2.5.15) holds for any $t \in [t_0, T)$. The proof of Theorem 2.5.1 is complete.

If the solutions $x(t)$ and $y(t)$ of inequalities (2.5.8) and (2.5.9), respectively, are continuously differentiable, then we give the following modified version of Theorem 2.5.1. The details of the proof which makes use of Lemma 2.5.1 and a continuous version of a general comparison principle proved in [16] are left to the reader.

Theorem 2.5.2

Assume $p_i(t)$, $\tau_i(t)$ $(1 \leq i \leq n)$ and $q_j(t)$, $\sigma_j(t)$ $(1 \leq j \leq m)$ are continuous and nonnegative functions. Suppose further that $x(t)$ is a positive solution of (2.5.8) and $y(t)$ is a solution of (2.5.9) such that $y(t)$ is positive and conditions (2.5.10)–(2.5.14) hold. Then condition (2.5.15) holds on $[t_0, T)$.

Remark 2.5.1 In the special case where $q_1(t) = \cdots = q_n(t) = 0$ on $[t_0, T)$, that is, when the inequalities investigated reduce to the delay case, Theorem 2.5.2 becomes precisely a result shown in [28].

Corollary 2.5.1 *Assume p_i, $\tau_i (1 \leq i \leq n)$, q_j, $\sigma_j (1 \leq j \leq m)$ are given positive numbers and $-\infty < t_0 \leq T < \infty$, moreover*

$$\tau = \max_{1 \leq i \leq n} \tau_i, \quad \sigma = \max_{1 \leq j \leq m} \sigma_j, \quad \gamma = \max\{\tau, 0\}.$$

Suppose further that the function $y \in C([t_0 - \gamma, T]; \mathbb{R})$ is such that $y' \in BM([t_0 - \sigma, T]; \mathbb{R})$, moveover,

$$y'(t) \geq \sum_{j=1}^{m} q_j y'(t - \sigma_j) - \sum_{i=1}^{n} p_i y(t - \tau_i), \quad \text{a.e. on } [t_0, T] \quad (2.5.24)$$

and

$$y(t) = d, \quad t_0 - \gamma \leq t \leq t_0 \quad (2.5.25)$$

where $d > 0$ is a constant.

If there exists a function $x \in C([t_0 - \gamma, T]; (0, \infty))$ such that

$$x(t) \geq x(t_0), \quad (t_0 - \gamma \leq t \leq t_0), \quad x' \in BM([t_0, T]; \mathbb{R}_+)$$

and

$$x'(t) \leq \sum_{j=1}^{m} q_j x'(t - \sigma_j) - \sum_{i=1}^{n} p_i x(t - \tau_i), \quad \text{a.e. on } [t_0, T] \quad (2.5.26)$$

then

$$y(t) > 0, \quad \text{a.e. on } [t_0, T]. \quad (2.5.27)$$

Proof

Let δ be an arbitrary fixed real number and define $x_\delta(t)$ by $x_\delta(t) = x(t) - \delta$, $t \geq t_0 - \gamma$. Then $x_\delta'(t) = x'(t) < 0$, $(t_0 - \sigma \leq t \leq T)$, and from equation (2.5.26) it follows that

$$x_\delta'(t) \leq \sum_{j=1}^{m} q_j x_\delta'(t - \sigma_j) - \sum_{i=1}^{n} p_i x_\delta(t - \tau_i) - \delta \sum_{i=1}^{n} p_i$$

for a.e. $t \in [t_0, T]$.
  Therefore,

$$x_\delta'(t) < \sum_{j=1}^{m} q_j x_\delta'(t - \sigma_j) - \sum_{i=1}^{n} p_i x_\delta(t - \tau_i), \quad \text{a.e. on } [t_0, T] \quad (2.5.28)$$

for any fixed $\delta > 0$.
  For any fixed $\delta \in (0, x(t_0))$, $x_\delta(t) > 0$, $t_0 - \gamma \leq t \leq t_0$, and

$$\frac{x_\delta'(t)}{x_\delta(t)} \leq \frac{y'(t)}{y(t)} = 0, \quad t_0 - \sigma \leq t \leq t_0$$

moreover,

$$\frac{d}{x_\delta(t)} = \frac{y(t)}{x_\delta(t)} \leq \frac{d}{x_\delta(t_0)} = \frac{y(t_0)}{x_\delta(t_0)}, \quad t_0 - \gamma \leq t \leq t_0.$$

Now, we show that $y(t) > 0$ on $t_0 \leq t \leq T$. Otherwise, there would exist $T_1 \in (t_0, T]$ such that

$$y(t) > 0, \quad t_0 \leq t < T_1 \quad \text{and} \quad y(T_1) = 0.$$

But in that case for fixed

$$\delta \in \left(0, \; \min_{t_0 \leq t \leq T_1} x(t) \right)$$

we have $x_\delta(t) > 0$ on $[t_0 - \gamma, T_1]$.

Thus in virtue of Theorem 2.5.1 we obtain

$$\frac{y(t)}{x_\delta(t)} > \frac{y(t_0)}{x_\delta(t_0)} > 0, \quad t_0 \leq t < T_1$$

that is $y(T_1) > 0$ which contradicts the definition of T_1. Therefore, $y(t) > 0$ on $[t_0, T_1]$ and the proof is complete.

Corollary 2.5.2 *Assume p_i, τ_i $(1 \leq i \leq n)$, and q_j, σ_j $(1 \leq j \leq m)$ are given positive numbers. Assume further that inequality (2.5.26) has an absolutely continuous solution on $[t_0 - \gamma, \infty)$ such that $x(t) > 0$ $(t_0 \leq t < \infty)$, and $x'(t) \leq 0$ (a.e. on $t_0 - \gamma \leq t < \infty$). Then equation*

$$y'(t) = \sum_{j=1}^{m} q_j y'(t - \sigma_j) - \sum_{i=1}^{n} p_i y(t - \tau_i) \tag{2.5.29}$$

has a positive solution on $[t_0 - \gamma, \infty)$.

The proof of the corollary follows from Corollary 2.5.1 in a direct way.

Remark 2.5.2 Corollary 2.5.2 extends to the neutral case a previous result of Ladas and Stavroulakis [105] for delay equations.

The next part is devoted to the study of the oscillatory properties of linear neutral functional differential equations and inequalities via their characteristic equations.

Consider the delay differential equation

$$x'(t) = - \sum_{i=0}^{n} p_i x(t - \tau_i) \tag{2.5.30}$$

where p_i and $\tau_i > 0$ $(0 \leq i \leq n)$ are given constants. In that case, it is known that equation (2.5.30) has a nonoscillating solution if and only if the equation

$$\lambda = \sum_{i=0}^{n} p_i \exp(\lambda \tau_i) \tag{2.5.31}$$

has a real root [6].

Motivated by the above result, a conjecture can be set for the functional equation of neutral type

$$x'(t) = \sum_{j=1}^{m} q_j x'(t - \sigma_j) - \sum_{i=0}^{n} p_i x(t - \tau_i) \qquad (2.5.32)$$

where p_i, $\tau_i > 0$ $(0 \leq i \leq n)$ and q_j, $\sigma_j > 0$ $(1 \leq j \leq m)$ are given constants.

Theorem 2.5.3

Assume p_i, τ_i $(0 \leq i \leq n)$, q_j, σ_j $(1 \leq j \leq m)$ *are positive constants,*

$$\tau = \max_{0 \leq i \leq n} \tau_i, \quad \sigma = \max_{1 \leq j \leq m} \sigma_j, \quad \gamma = \max\{\tau, \sigma\}$$

and

$$\sum_{j=1}^{m} q_j < 1. \qquad (2.5.33)$$

Then each solution of equation (2.5.32) oscillates if and only if the equation

$$\lambda = \lambda \sum_{j=1}^{m} q_j \exp(\lambda \sigma_j) + \sum_{i=0}^{n} p_i \exp(\lambda \tau_i) \qquad (2.5.34)$$

has no real root.

We shall use the following statements to prove the above theorem.

Lemma 2.5.2

Suppose that $a > 0$, $\delta > 0$ *and* $T > -\infty$ *are given real numbers. If* $\alpha : [T - \delta, \infty) \to \mathbb{R}_+$ *is a continuous function and*

$$\alpha(t) \geq a \exp\left(\int_{t-\delta}^{t} \alpha(s)\, ds \right), \quad t \geq T \qquad (2.5.35)$$

then

$$\liminf_{t \to \infty} \alpha(t) < \infty. \qquad (2.5.36)$$

This lemma was proved in [60], therefore we will not give the details here (see also, for instance, [61], [64]).

Lemma 2.5.3

Assume that the conditions of Theorem 2.5.3 hold. If equation (2.5.32) has an eventually positive (negative) solution, then there exists $T > 0$ such that equation (2.5.32) has a continuously differentiable positive and monotone decreasing solution on $[T - \gamma, \infty)$.

Proof

It suffices to consider the case when equation (2.5.32) has an eventually positive solution, because if $x(t)$ is a solution, then $-x(t)$ is also a solution.

Assume $x(t)$ is an eventually positive solution of equation (2.5.32) on $[T - \gamma, \infty)$ and define $y(t)$ to be the function

$$y(t) = \int_{T_1}^{t} x(s)\, ds - c, \quad t \geq T_1 - \gamma$$

where $T_1 \geq 0$ is such that $x(t) > 0$, $t \geq T_1 - \gamma$,

$$c = c_1 \left(\sum_{i=0}^{n} p_i \right)^{-1}$$

and   (2.5.37)

$$c_1 = x(T_1) - \sum_{j=1}^{m} q_j x(T_1 - \tau_j) - \sum_{i=0}^{n} p_i \int_{T_1 - \tau_i}^{T_1} x(u)\, du.$$

Integrate both sides of equation (2.5.32) from T_1 to t, we have

$$x(t) = x(T_1) + \sum_{j=1}^{m} q_j [x(t - \sigma_j) - x(T_1 - \sigma_j)] - \sum_{i=0}^{n} p_i \int_{T_1 - \tau_i}^{t - \tau_i} x(s)\, ds.$$

Using the definition of c_1, we obtain

$$x(t) = \sum_{j=1}^{m} q_j x(t - \sigma_j) + c_1 - \sum_{i=0}^{n} p_i \int_{T_1}^{t - \tau_i} x(s)\, ds$$

that is

$$x(t) = \sum_{j=1}^{m} q_j x(t - \sigma_j) - \sum_{i=0}^{n} p_i \left[\int_{T_1}^{t - \tau_i} x(s)\, ds - c \right].$$

But from the definition of y we have

$$y'(t) = \sum_{j=1}^{m} q_j y'(t - \sigma_j) - \sum_{i=0}^{n} p_i y(t - \tau_i), \quad t \geq T_1 + \gamma \quad (2.5.38)$$

which means that $y(t)$ is a continuously differentiable solution of equation (2.5.32) on $[T_1, \infty)$. Moreover, $y(t)$ is monotone increasing on $[T_1, \infty)$ since $y'(t) = x(t) > 0$.

This implies that $y(t)$ does not oscillate on $[T_1 - \gamma, \infty)$, that is, $y(t)$ is eventually positive or eventually negative. We show that $y(t)$ cannot be eventually positive under our conditions. Indeed, if $y(t)$ is eventually positive, then there exists $T_2 \geq T_1$ such that $y(t) > y(T_2 - \tau) > 0$, $t \geq T_2 - \tau$, and thus

$$\sum_{i=0}^{n} p_i y(t - \tau_i) > \sum_{i=0}^{n} p_i y(T_2 - \tau) \equiv m > 0, \quad t > T_2.$$

Moreover, $y(t)$ satisfies equation (2.5.38); hence it follows that

$$y'(t) < \sum_{j=1}^{m} q_j y'(t - \sigma_j) - m \leq \max_{0 \leq s \leq \sigma} y'(t - s) - m, \quad t \geq T_2 \quad (2.5.39)$$

since $\sum_{j=1}^{m} q_j < 1$. But relation (2.5.39) implies

$$y'(t) < \max_{0 \leq s \leq \sigma} y'(T_2 - s), \quad t \geq T_2$$

and thus

$$0 \leq \limsup_{t \to +\infty} y'(t) \leq \limsup_{t \to +\infty} y'(t) - m < \limsup_{t \to \infty} y'(t) < \infty$$

which is a contradiction. Hence $y(t)$ cannot be eventually positive which implies that there exists $T_3 > 0$ such that $z_1(t) = -z(t + T_3)$ is positive and its derivative is negative on $[T - \gamma, \infty)$.

On the other hand, equation (2.5.32) is autonomous and homogeneous, thus $z_1(t)$ is a solution of equation (2.5.32) on $[T - \gamma, \infty)$. The proof of the lemma is complete.

Proof of Theorem 2.5.3

It is clear that a positive root λ of equation (2.5.34) leads to a solution $x(t) = \exp(-\lambda t)$ which is positive.

Now we discuss the case when equation (2.5.32) has a nonoscillating solution on $\mathbb{R}_+$, say $x(t)$. Then by virtue of Lemma 2.5.3 we have that equation (2.5.32) has a continuously differentiable and decreasing solution $y(t)$ on $[-\gamma, \infty)$.

Thus $y(t)$ can be written in the form

$$y(t) = y_0 \exp\left(-\int_{t_0}^{t} \alpha(u)\,du\right), \quad t \ge -\gamma$$

where $x_0 = x(-\gamma)$ and

$$\alpha(t) = -\frac{y'(t)}{y(t)}$$

is a continuous and nonnegative function on $[-\gamma, \infty)$. From equation (2.5.32) it is easy to see that $\alpha(t)$ satisfies the equation

$$\alpha(t) = \sum_{j=1}^{m} q_j \alpha(t-\sigma_j) \exp\left(\int_{t-\sigma_j}^{t} \alpha(u)\,du\right) + \sum_{i=0}^{n} p_i \exp\left(\int_{t-\tau_i}^{t} \alpha(u)\,du\right)$$

$$(2.5.40)$$

for any $t \ge 0$.

Therefore,

$$\alpha(t) \ge a \exp\left(\int_{t-\delta}^{t} \alpha(u)\,du\right), \quad t \ge 0$$

where

$$a = \min_{0 \le i \le n} p_i > 0, \quad \delta = \min_{0 \le i \le n} \tau_i.$$

Thus from Lemma 2.5.2 we have

$$a = \lim_{t \to +\infty} \inf \alpha(t) < \infty.$$

In that case, equation (2.5.40) yields

$$a \geq \sum_{j=1}^{m} q_j \, a \exp(a\sigma_j) + \sum_{i=0}^{n} p_i \exp(a\tau_i).$$

From the last inequality one can see that equation (2.5.34) has a positive root. The proof of the theorem is complete.

Remark 2.5.3   When $m = 1$, Theorem 2.5.3 reduces to a known result of Kulenovic *et al* [94], but the next theorem on neutral equations is not known to us in any literature.

Theorem 2.5.4

Assume all conditions of Theorem 2.5.3 hold. Then the inequality

$$x'(t) \leq \sum_{j=1}^{m} q_j x'(t - \sigma_j) - \sum_{i=0}^{n} p_i x(t - \tau_i) \qquad (2.5.41)$$

has a positive and monotone decreasing solution on an interval $[T - \gamma, \infty)$ if and only if equation (2.5.34) has a real root.

Proof

The theorem is a simple consequence of Corollary 2.5.2 and Theorem 2.5.3; therefore, we omit the details.

In the above theorems we investigated linear equations as well as inequalities. Now we apply these results to investigate the oscillatory properties of solutions of the time-delayed logistic equation with neutral terms.

The present results are given via the linear approximations of the nonlinear equations investigated, similar to the stability analysis of perturbed linear systems.

Now, consider the following general nonlinear equation

$$x'(t) = \sum_{j=1}^{m} q_j x'(t - \sigma_j) - \sum_{i=0}^{n} p_i x(t - \tau_i)$$

$$- g(t, x(t - \tau_0), \dots, x(t - \tau_n)) \qquad (2.5.42)$$

where

H2.5.3 p_i, τ_i $(0 \leq i \leq n)$ and q_j, σ_j $(1 \leq j \leq m)$ are given nonnegative constants,

$$\sum_{j=1}^{m} q_j < 1, \quad \tau = \max_{0 \leq i \leq n} \tau_i, \quad \sigma = \max_{1 \leq j \leq m} \sigma_j \quad \text{and} \quad \gamma = \max\{\tau, \sigma\};$$

H2.5.4 $g : \mathbb{R}_+^{n+2} \to \mathbb{R}_+$ is a continuous function such that for any $\varepsilon > 0$ there exists $\delta_\varepsilon > 0$ such that

$$g(t, x_0, x_1, \ldots, x_n) \leq \varepsilon \sum_{i=0}^{n} p_i x_i \tag{2.5.43}$$

for any $t \geq 0$ and $x_i \in (0, \delta_\varepsilon)$, $0 \leq i \leq n$.

Remark 2.5.4 Without mentioning explicitly, we shall assume that each solution of the equations considered exists on $[-\gamma, \infty)$.

Theorem 2.5.5

Let conditions H2.5.3 and H2.5.4 hold. Then
  (a) if the equation

$$\lambda = \lambda \sum_{j=1}^{m} q_j \exp(\lambda \sigma_j) + \sum_{i=0}^{n} p_i \exp(\lambda \tau_i) \tag{2.5.44}$$

has a real root, then equation (2.5.42) has a positive solution on $[-\gamma, \infty)$;
  (b) if equation (2.5.42) has an eventually positive and eventually decreasing solution on $[-\gamma, \infty)$, then equation (2.5.44) has a real root.

Proof

Assume equation (2.5.44) has a real root, say λ_0. Then from $M = \sum_{j=1}^{m} q_j < 1$ it follows that $\lambda_0 > 0$. But it is easy to see that in that case there exists $\varepsilon \in (0, 1)$ such that the equation

$$\lambda = \lambda \sum_{j=1}^{m} q_j \exp(\lambda \sigma_j) + (1 - \varepsilon) \sum_{i=0}^{n} p_i \exp(\lambda \tau_i) \tag{2.5.45}$$

has a positive root, say λ_ε. Therefore

$$x_\varepsilon(t) = \exp(-\lambda_\varepsilon t) \tag{2.5.46}$$

is a positive and monotone decreasing solution of

$$x'(t) - \sum_{j=1}^{m} q_j x'(t - \sigma_j) = -(1 - \varepsilon) \sum_{i=0}^{n} p_i x(t - \tau_i). \qquad (2.5.47)$$

Let $d \in (0, \delta_\varepsilon)$ be a fixed constant and consider a solution $y: [-\gamma, \infty) \to \mathbb{R}$ of equation (2.5.42) such that

$$y(t) = d, \quad t_0 - \gamma \leq t \leq t_0.$$

We shall show that $y(t) > 0$ on $[-\gamma, \infty)$. If it is not true, then there exists $T > 0$ such that

$$y(t) > 0, \quad -\gamma \leq t < T \quad \text{and} \quad y(T) = 0.$$

But $y(t)$ is a solution of equation (2.5.42) and y is a nonnegative function, therefore, we have

$$y'(t) > \sum_{j=1}^{m} q_j y'(t - \sigma_j) - (1 - \varepsilon) \sum_{i=0}^{n} p_i y(t - \tau_i), \quad 0 \leq t \leq T. \qquad (2.5.48)$$

In view of equation (2.5.47) and (2.5.48), from Corollary 2.5.1 it follows that $y(t) > 0$ for any $t \in [t_0, T]$, which is a contradiction. Therefore, $y(t)$ is a positive solution of equation (2.5.42) on $[-\gamma, \infty)$.

Now, we show that if equation (2.5.42) has an eventually positive or an eventually decreasing solution on $[-\gamma, \infty)$, say $x(t)$, then equation (2.5.44) has a real root. Namely, in that case there exists $T > 0$ such that $x(t) > 0$ and $x'(t) < 0$, $t \geq T - \gamma$; moreover, equation (2.5.42) implies

$$x'(t) \leq \sum_{j=1}^{m} q_j x'(t - \sigma_j) - \sum_{i=0}^{n} p_i x(t - \tau_i), \quad t \geq T.$$

By Theorem 2.5.3 this implies that equation (2.5.44) has a real root. The proof of the theorem is complete.

Now, by Theorem 2.5.5 we obtain the following result.

Theorem 2.5.6

Assume p_i, τ_i $(0 \leq i \leq n)$ are given positive constants and $g: \mathbb{R}_+^{n+2} \to \mathbb{R}_+$ satisfies condition H2.5.4. Then the equation

$$x'(t) = -\sum_{i=0}^{n} p_i x(t - \tau_i) - g(t, x(t - \tau_0), \ldots, x(t - \tau_n)) \qquad (2.5.49)$$

has an eventually positive solution on $[-\gamma, \infty)$ *if and only if the equation*

$$\lambda = \sum_{i=0}^{n} p_i \exp(\lambda \tau_i) \qquad (2.5.50)$$

has a real root.

Proof

Equations (2.5.49) and (2.5.50) are special cases of equations (2.5.42) and (2.5.44), respectively, where $q_j = 0$, $1 \leq j \leq m$. On the other hand, if equation (2.5.49) has a positive solution $x(t)$, then $x'(t)$ is eventually negative. Therefore from Theorem 2.5.5 it follows that equation (2.5.49) has a positive solution if and only if equation (2.5.50) has a real root, which completes the proof.

Now, consider the delay logistic equation in the following form:

$$x'(t) = \sum_{j=1}^{m} q_j x'(t - \sigma_j) + rx(t)\left[1 - \frac{1}{K}\sum_{i=1}^{n} p_i x(t - \tau_i)\right] \quad (2.5.51)$$

where we assume that:

H2.5.5 The weight constants p_i and the delays τ_i are positive for any $1 \leq i \leq n$ and

$$\sum_{i=1}^{n} p_i = 1, \quad \tau = \max_{1 \leq i \leq n} \tau_i. \qquad (2.5.52)$$

H2.5.6 The intrinsic growth rate r and the carrying capacity K are positive real numbers.

H2.5.7 The speed weight constants q_j and the delays σ_j are nonnegative for any $1 \leq j \leq m$ and

$$\sum_{j=1}^{m} q_j < 1. \qquad (2.5.53)$$

Remark 2.5.5 Equation (2.5.51) was introduced in [59] where some qualitative properties of its solution were also studied.

It is easy to see that equation (2.5.51) has only one positive steady state solution K. Now as we are interested in the oscillatory properties of the solutions of equation (2.5.51) around the steady state, we make the following transformation

$$y(t) = x(t) - K, \quad -\gamma \leq t < \infty.$$

Then it is easy to see that $y(t)$ is a solution of the equation

$$y'(t) = \sum_{j=1}^{m} q_j y'(t - \sigma_j) - r \sum_{i=1}^{n} p_i y(t - \tau_i) - \frac{r}{K} y(t) \sum_{i=1}^{n} p_i y(t - \tau_i)$$

$$(2.5.54)$$

on $[-\gamma, \infty)$.

Therefore from Theorem 2.5.5 and Theorem 2.5.6 the following results are immediately obtained.

Theorem 2.5.7

Let H2.5.5–H2.5.7 hold. Then
  (a) if the equation

$$\lambda = \lambda \sum_{j=1}^{m} q_j \exp(\lambda \sigma_j) + \sum_{i=1}^{n} p_i \exp(\lambda \tau_i) \qquad (2.5.55)$$

has a real root, then equation (2.5.51) has a solution $x: [-\gamma, \infty) \to \mathbb{R}$ *such that* $y(t) = x(t) - K > 0$, $-\gamma \le t < \infty$;
  (b) If equation (2.5.51) has a solution $x: [-\gamma, \infty) \to \mathbb{R}$ *such that* $y(t) = x(t) - K$ *is eventually positive and decreasing, then equation (2.5.55) has a real root.*

Theorem 2.5.8

Let H2.5.5 and H2.5.6 hold. Then the delay logistic equation without neutral terms

$$x'(t) = rx(t) \left[1 - \frac{1}{K} \sum_{i=1}^{n} p_i x(t - \tau_i) \right] \qquad (2.5.56)$$

has a solution $x: [-\gamma, \infty) \to \mathbb{R}$ *such that* $y(t) = x(t) - K$ *is eventually positive if and only if the equation*

$$\lambda = r \sum_{i=1}^{n} p_i \exp(\lambda \tau_i) \qquad (2.5.57)$$

has a real root.

The advantage of Theorems 2.5.7 and 2.5.8 is that many necessary and sufficient conditions are known for equations (2.5.55) and (2.5.57) to have only nonreal roots as well as to have a real root (see, for example, [105] and the references cited therein). For example, using some known results (see, for example, [6], [102], [161]), Theorem 2.5.7 implies the following:

Corollary 2.5.3 *Assume* $r > 0$, $K > 0$, $p_i > 0$ *and* $\tau_i > 0$, $(1 \leq i \leq n)$ *are given constants,*

$$\sum_{i=1}^{n} p_i = 1, \quad \tau = \max_{1 \leq i \leq n} \tau_i.$$

Then

(a) *if*

$$e \sum_{i=1}^{n} p_i \tau_i > 1$$

each solution of equation (2.5.54) oscillates about the steady state K;

(b) *if*

$$e \tau \sum_{i=1}^{n} p_i \leq 1$$

then there exists a solution $x : [-\gamma, \infty) \to \mathbb{R}$ *of equation (2.5.54) such that* $x(t) - K$ *is eventually positive.*

Remark 2.5.6 Corollary 2.5.3 was proved by Gopalsamy in [37], [38] using a different technique. Equation (2.5.54) was investigated in a recent paper, and also by Kulenovic *et al* [95], but from their results the sharp Theorem 2.5.7 does not follow. Theorem 2.5.6 has not been seen in any earlier publications.

Remark 2.5.7 Using this technique, one can generalize Theorem 2.5.6 for the case where $p_i = p_i(t)$ and $\tau_i = \tau_i(t)$ are continuous functions. But in that case we have to use the linear differential equation

$$x'(t) = -\sum_{i=1}^{n} p_i(t) x(t - \tau_i(t))$$

without equation (2.5.50). From this generalized form of Theorem 2.5.6 there follow some interesting results for the time-dependent delay logistic equation, that is where $r = r(t)$ and $K = K(t)$.

2.6   Notes and comments to Chapter 2

The results of Section 2.1 are due to Grammatikopoulos, Ladas and Sficas [51]. Oscillatory properties of the solutions of neutral delay differential equations of the form

$$\frac{\mathrm{d}}{\mathrm{d}t}[y(t) + py(t - \tau)] + Q(t)y(t - \sigma) = 0, \quad t \geqq t_0$$

where $p \in \mathbb{R}$, were investigated by Ladas and Sficas [101], Grammatikopoulos, Grove and Ladas [47] and Zhang [162]. Generalizations of some results of Section 2.1 for equation (2.1.1) were obtained by Grove *et al* [56]. The same authors obtained comparison theorems of Myshkis's type for neutral equations [55]. The asymptotic behaviour of the solutions of neutral equations was investigated by Ntouyas and Sficas [129] and Ladas and Sficas [99]. The question of forced oscillations was considered by Erbe and Zhang [30]. Sufficient conditions for oscillation of neutral differential equations with variable delay were obtained by Györi [64]. Oscillatory properties and asymptotic behaviour of the solutions of neutral equations were also investigated by Ruan [135], [136]. Theorems on the distribution of the zeros of the solutions of first order neutral equations were obtained by Domshlak [22], Domshlak and Sheikhzamanova [23] and Aliev [2].

The results of Section 2.2 are due to Sficas and Stavroulakis [139]. A generalization of their result was obtained in the works of Grove and Ladas [57], Kulenović *et al* [94], Grammatikopoulos, Sficas and Stavroulakis [52], Farrell [32], Zhanyuan [163] and Grammatikopoulos and Stavroulakis [53], [54]. We shall especially note the result of Grammatikopoulos and Stavroulakis [54] where a necessary and sufficient condition is obtained for oscillation of neutral differential equations of the form

$$\frac{\mathrm{d}}{\mathrm{d}t}\left[x(t) + \sum_I p_i x(t - \tau_i)\right] + \sum_K q_k x(t - \sigma_k) = 0$$

where p_i, τ_i, q_k, $\sigma_k \in \mathbb{R}$. Sufficient conditions for oscillation and asymptotic behaviour of the solutions of neutral differential equations with constant coefficients were obtained by Ladas and Sficas [101], Grammatikopoulos, Grove and Ladas [45], [47], Zhang [162] and Partheniadis [130]. Sufficient conditions for oscillation of neutral equations of mixed type were obtained by Stavroulakis [146].

The results of Section 2.3 are due to Bainov *et al* [8]. Sufficient conditions for oscillation of the solutions of neutral differential equations with distributed delay were obtained in the works of Bainov *et al* [11] and Györi [64].

The results of Section 2.4 are due to Jaroš and Kusano [76]. Sufficient conditions for oscillation of the solutions of nonlinear first order neutral differential equations were obtained by Ivanov [72], Ivanov and Kusano [73], [74] and Ladas [97]. The asymptotic behaviour of the nonoscillating solutions of neutral differential equations with delay is investigated by Graef *et al* [44]. Asymptotic properties of the solutions of neutral equations were also investigated by Arino and Bourad [4].

The results of Section 2.5 are due to Györi [63]. Oscillatory properties of the solutions of the neutral delay logistic equation were investigated in the works of Gopalsamy and Zhang [39] and Györi [62].

3

Second order neutral ordinary differential equations

3.1 Second order linear differential equations

In this section the oscillatory properties and asymptotic behaviour of the solutions of neutral differential equations of the form

$$\frac{d^2}{dt^2}[y(t) + P(t)y(t - \tau)] + Q(t)y(t - \sigma) = 0, \quad t \geq t_0 \quad (3.1.1)$$

are investigated, where $P(t)$, $Q(t) \in C([t_0, \infty); \mathbb{R})$ and the delays τ and σ are nonnegative real numbers.

Let $\varphi(t) \in C([t_0 - \rho, t_0]; \mathbb{R})$, where $\rho = \max\{\tau, \sigma\}$, is a given function and let z_1 be a given constant.

Definition 3.1.1 The function $y(t) \in C([t_0 - \rho, \infty); \mathbb{R})$ is said to be a solution of equation (3.1.1) if

$$y(t) = \varphi(t), \quad t \in [t_0 - \rho, t_0]$$

$$\frac{d}{dt}[y(t) + P(t)y(t - \tau)]|_{t=t_0} = z_1$$

the function $y(t) + P(t)y(t - \tau)$ is twice differentiable for $t \geq t_0$ and $y(t)$ satisfies equation (3.1.1) for $t \geq t_0$.

We shall note that theorems of existence and uniqueness of the solution of neutral differential equations were obtained by Driver [26], [27], Bellman and Cooke [16] and Hale [66].

First we shall consider the asymptotic behaviour of the nonoscillating solutions of equation (3.1.1).

We shall say that conditions (H3.1) are met if the following conditions hold:

H3.1.1 $P(t) \in C([t_0, \infty); \mathbb{R})$, $p_1 \le P(t) \le p_2$ for $t \in [t_0, \infty)$, where p_1 and p_2 are constants.

H3.1.2 $Q(t) \in C([t_0, \infty); \mathbb{R})$, $Q(t) \ge q = \text{const.} > 0$ for $t \in [t_0, \infty)$.

Let $y(t)$ be a solution of equation (3.1.1). Set

$$z(t) = y(t) + P(t)y(t - \tau).$$

The following lemma describes some asymptotic properties of the function $z(t)$ when $y(t)$ is a nonoscillating solution of equation (3.1.1).

Lemma 3.1.1

Assume conditions H3.1.1 and H3.1.2 fulfilled. Let $y(t)$ be an eventually positive solution of equation (3.1.1). Set

$$z(t) = y(t) + P(t)y(t - \tau). \tag{3.1.2}$$

Then the following statements are true:

(a) The functions $z(t)$ and $z'(t)$ are strictly monotone and either

$$\lim_{t \to \infty} z(t) = \lim_{t \to \infty} z'(t) = -\infty \tag{3.1.3}$$

or

$$\lim_{t \to \infty} z(t) = \lim_{t \to \infty} z'(t) = 0, \quad z(t) < 0 \quad \text{and} \quad z'(t) > 0. \tag{3.1.4}$$

In particular, $z(t)$ is always negative.

(b) Assume that $p_1 \ge -1$. Then (3.1.4) holds. In particular, $z(t)$ is bounded.

Proof

(a) From equation (3.1.1) we have

$$z''(t) = -Q(t)y(t - \sigma) \le -qy(t - \sigma) < 0 \tag{3.1.5}$$

which implies that $z'(t)$ is a strictly decreasing function of t and so $z(t)$ is a strictly monotone function. From the above observation it follows that either

$$\lim_{t \to \infty} z(t) = \lim_{t \to \infty} z'(t) = -\infty$$

or

$$\lim_{t \to \infty} z'(t) \equiv l \quad \text{is finite.} \tag{3.1.6}$$

Assume that (3.1.6) holds. Then, integrating both sides of (3.1.5) from t_1 to t, with t_1 large enough, and letting $t \to \infty$, we find

$$\int_{t_0}^{\infty} qy(t - \sigma)\, ds \leqq z'(t_1) - l$$

which implies that $y \in L_1[t_1, \infty)$ and so $z \in L_1[t_1, \infty)$. Since $z(t)$ is monotone, it follows that

$$\lim_{t \to \infty} z(t) = 0 \tag{3.1.7}$$

and therefore $l = 0$. Finally, by (3.1.7) and (3.1.6) with $l = 0$ and the decreasing nature of $z'(t)$ we conclude that

$$z(t) < 0 \quad \text{and} \quad z'(t) > 0.$$

(*b*) If (3.1.4) were false, then from (3.1.3) it would follow that

$$\lim_{t \to \infty} z(t) = -\infty. \tag{3.1.8}$$

Using the fact that $p_1 \geqq -1$ and $z(t) < 0$, we find

$$y(t) < -P(t)y(t - \tau) \leqq -p_1 y(t - \tau) \leqq y(t - \tau)$$

which implies that $y(t)$ is bounded. This contradicts (3.1.8) and proves that (3.1.4) is satisfied. The proof of the lemma is complete.

Using the asymptotic properties of the function $z(t)$, we now prove the following result about the asymptotic behaviour of the nonoscillating solutions of equation (3.1.1).

Theorem 3.1.1

Consider the neutral delay differential equation (3.1.1) and assume conditions H3.1.1 and H3.1.2 fulfilled,

$$-1 < p_1 \leqq p_2 < 0. \tag{3.1.9}$$

Then each nonoscillating solution $y(t)$ of equation $(3.1.1)$ tends to zero as $t \to \infty$.

Proof

As the negative of a solution of equation $(3.1.1)$ is also a solution of the same equation, it suffices to prove the theorem for an eventually positive solution $y(t)$ of equation $(3.1.1)$.

Set

$$z(t) = y(t) + P(t)y(t - \tau).$$

Then Lemma 3.1.1 (b) implies that $(3.1.4)$ holds. In particular,

$$z(t) < 0$$

and hence

$$y(t) < -P(t)y(t - \tau) < y(t - \tau).$$

Therefore, $y(t)$ is a bounded function. Assume, for the sake of contradiction, that

$$\limsup_{t \to \infty} y(t) \equiv s > 0. \tag{3.1.10}$$

Let $\{t_n\}$ be a sequence of points such that

$$\lim_{n \to \infty} t_n = \infty \quad \text{and} \quad \lim_{n \to \infty} y(t_n) = s.$$

Then, for n large enough,

$$z(t_n) = y(t_n) + P(t_n)y(t_n - \tau) \geq y(t_n) + p_1 y(t_n - \tau)$$

and so

$$\limsup_{n \to \infty} y(t_n - \tau) \geq s/(-p_1) > s.$$

This contradicts $(3.1.10)$ and the proof is complete.

The following example illustrates that, if condition H3.1.2 of Theorem 3.1.1 is violated, the result may not be true.

Example 3.1.1   In the neutral delay differential equation

$$\frac{d^2}{dt^2}[y(t) + (-\tfrac{1}{2} + (t-1)^{-\frac{1}{2}})y(t-1)]$$

$$+ \tfrac{1}{4}(t-2)^{-\frac{1}{2}}(t^{-\frac{3}{2}} - \tfrac{1}{2}(t-1)^{-\frac{3}{2}})y(t-2) = 0, \quad t \geq 2$$

all conditions of Theorem 3.1.1, except for H3.1.2, are satisfied. Note that the function $y(t) = \sqrt{t}$ is a solution with $\lim_{t \to \infty} y(t) = \infty$.

In the subsequent theorems sufficient conditions are given for oscillation of the solutions of equation (3.1.1).

The following lemma, which will be utilized in the proofs of some of the theorems, is due to Ladas and Stavroulakis [104].

Lemma 3.1.2 [104]

Assume that r and μ are positive constants such that

$$r^{\frac{1}{2}} \frac{\mu}{2} > \frac{1}{e}.$$

Then the inequality

$$x''(t) - rx(t - \mu) \leq 0$$

has no eventually negative bounded solution.
    Let $y(t)$ be a solution of equation (3.1.1). Set

$$z(t) = y(t) + P(t)y(t - \tau).$$

Then a direct substitution shows that $z(t)$ is a twice continuously differentiable solution of the neutral delay differential equation

$$z''(t) + R(t)z''(t - \tau) + Q(t)z(t - \sigma) = 0, \quad t \geq t_0 \qquad (3.1.11)$$

where

$$R(t) = P(t - \sigma)\frac{Q(t)}{Q(t - \tau)}.$$

Theorem 3.1.2

Consider the neutral delay differential equation (3.1.1) and assume conditions H3.1.1 and H3.1.2 fulfilled. Furthermore, assume that $P(t)$ is not eventually negative. Then each solution of equation (3.1.1) oscillates.

Proof

Assume, for the sake of contradiction, that there is an eventually positive solution $y(t)$ of equation (3.1.1). Set

$$z(t) = y(t) + P(t)y(t - \tau).$$

Then, eventually, $z(t)$ takes nonnegative values. However, Lemma 3.1.1 (a) implies that $z(t)$ is eventually negative. This is a contradiction and the proof is complete.

The example below illustrates Theorem 3.1.2.

Example 3.1.2 The neutral delay differential equation

$$\frac{d^2}{dt^2} [y(t) + (\tfrac{1}{2} + \sin t)y(t - 2\pi)]$$

$$+ (\tfrac{3}{2} + \sin t)y(t - 4\pi) = 0, \quad t \geq 0 \qquad (3.1.12)$$

satisfies the conditions of Theorem 3.1.2. Therefore each solution of equation (3.1.12) oscillates. For example $y(t) = \sin t(\tfrac{3}{2} + \sin t)^{-1}$ is an oscillating solution.

The following example illustrates that if condition H3.1.2 of Theorem 3.1.2 is violated, the result may not be true.

Example 3.1.3 For the neutral delay differential equation

$$\frac{d^2}{dt^2} [y(t) + (t - 1)^{-\frac{1}{2}} y(t - 1)] + \tfrac{1}{4}t^{-\frac{3}{2}}(t - 2)^{-\frac{1}{2}}y(t - 2) = 0, \quad t > 2$$

all conditions of Theorem 3.1.2, except for H3.1.2, hold. Note, however, that $y(t) = \sqrt{t}$ is a nonoscillating solution.

Theorem 3.1.3

Consider the neutral delay differential equation (3.1.1) and assume that condition H3.1.1 and H3.1.2 are satisfied with

$$-1 \leqq p_1 \leqq p_2 < 0. \tag{3.1.13}$$

Suppose also that there exists a positive constant r such that

$$\frac{Q(t)}{P(t + \tau - \sigma)} \leqq -r \tag{3.1.14}$$

and

$$r^{\frac{1}{2}} \frac{\sigma - \tau}{2} > \frac{1}{e}. \tag{3.1.15}$$

Then each solution of equation (3.1.1) oscillates.

Proof

Otherwise, there exists an eventually positive solution $y(t)$ of equation (3.1.1). Set

$$z(t) = y(t) + P(t)y(t - \tau).$$

Then $z(t)$ is a solution of equation (3.1.11). From equation (3.1.1) we have,

$$z''(t) < 0 \tag{3.1.16}$$

and, in view of (3.1.13), Lemma 3.1.1 (b) implies that $z(t)$ is eventually a negative bounded function. Using (3.1.16), equation (3.1.11) yields

$$R(t)z''(t - \tau) + Q(t)z(t - \sigma) > 0.$$

Hence, in view of (3.1.14), we find

$$z''(t) - rz(t - (\sigma - \tau)) < 0.$$

But, because of (3.1.15), Lemma 3.1.2 implies that it is impossible for this inequality to have an eventually negative bounded solution. This is a contradiction and the proof is complete.

Example 3.1.4 illustrates that if the condition $-1 \leqq p_1$ of Theorem 3.1.3 is violated, the result may not be true.

Example 3.1.4 For the neutral delay differential equation

$$\frac{d^2}{dt^2} [y(t) - (e^2 + e^{-t})y(t-1)] + e^2(e-1)y(t-2) = 0, \quad t \geqq 0$$

all conditions of Theorem 3.1.3, except for $-1 \leqq p_1$, are satisfied. Note that $y(t) = e^t$ is a nonoscillating solution of this equation.

Theorem 3.1.4

Consider the neutral delay differential equation (3.1.1) and assume conditions H3.1.1 and H3.1.2 fulfilled with

$$p_2 < 0. \tag{3.1.17}$$

Suppose also that there exists a positive constant r such that

$$\frac{Q(t)}{P(t + \tau - \sigma)} \leqq -r \tag{3.1.18}$$

and

$$r^{\frac{1}{2}} \frac{\sigma - \tau}{2} > \frac{1}{e}. \tag{3.1.19}$$

Then each bounded solution of equation (3.1.1) oscillates.

Proof

Otherwise, there exists an eventually positive bounded solution $y(t)$ of equation (3.1.1). Set

$$z(t) = y(t) + P(t)y(t - \tau).$$

Then $z(t)$ is a solution of equation (3.1.11). From equation (3.1.1) we have

$$z''(t) < 0 \tag{3.1.20}$$

and from Lemma 3.1.1 (a) we know that $z(t)$ is an eventually negative function. From condition H3.1.1 and since $y(t)$ is bounded, it follows that $z(t)$ is bounded. Using (3.1.20), equation (3.1.11) yields

$$R(t)z''(t - \tau) + Q(t)z(t - \sigma) > 0$$

and in view of (3.1.18) we obtain

$$z''(t) - rz(t - (\sigma - \tau)) < 0.$$

But, because of (3.1.19), Lemma 3.1.2 implies that it is impossible for this inequality to have an eventually negative bounded solution. This is a contradiction and the proof is complete.

It should be noted that Example 3.1.4 also demonstrates that under the conditions of Theorem 3.1.4, equation (3.1.1) may have unbounded non-oscillating solutions.

Example 3.1.5   The neutral delay differential equation

$$\frac{d^2}{dt^2}\left[y(t) - \frac{\sqrt{2}}{2} y\left(t - \frac{\pi}{4} \right) \right] + \frac{\sqrt{2}}{2} y\left(t - \frac{7\pi}{4} \right) = 0, \quad t \geq 0,$$

satisfies all conditions of Theorem 3.1.4. Therefore each bounded solution of this equation oscillates. For instance, $y(t) = \sin t$ is such a solution.

In Theorems 3.1.5 and 3.1.6 below condition H3.1.2 is not required.

Theorem 3.1.5

Consider the neutral delay differential equation (3.1.1). Assume the following conditions eventually fulfilled:

$$Q(t) \geq 0, \quad -1 \leq P(t) \leq 0 \tag{3.1.21}$$

and

$$\int_{t_0}^{\infty} Q(s)\, ds = \infty.$$

Then each unbounded solution of equation (3.1.1) oscillates.

Proof

Otherwise, there exists an eventually unbounded solution $y(t)$ of equation (3.1.1). Set

$$z(t) = y(t) + P(t)y(t - \tau).$$

Then

$$z''(t) = -Q(t)y(t - \sigma) \leq 0 \qquad (3.1.22)$$

and so $z'(t)$ is a decreasing function. We claim that

$$z(t) > 0. \qquad (3.1.23)$$

Otherwise

$$y(t) \leq -P(t)y(t - \tau) \leq y(t - \tau)$$

which implies that $y(t)$ is bounded. This contradiction shows that (3.1.23) is satisfied. We also claim that

$$z'(t) \geq 0 \quad \text{eventually.}$$

Otherwise,

$$z'(t) < 0 \quad \text{and} \quad z''(t) \leq 0 \quad \text{eventually}$$

which, as $t \to +\infty$, leads to a contradiction. The proof is complete.

$$\lim_{t \to \infty} z(t) = -\infty.$$

This contradicts (3.1.23). Clearly

$$y(t) > z(t)$$

and so (3.1.22) yields

$$z''(t) + Q(t)z(t - \sigma) \leq 0. \qquad (3.1.24)$$

Integrating (3.1.24) from t_1 to t, with t_1 large enough, we get

$$z'(t) - z'(t_1) + z(t_1 - \sigma) \int_{t_1}^{t} Q(s)\, ds \leq 0$$

which, as $t \to +\infty$, leads to a contradiction. The proof is complete.

The following example shows that under the conditions of Theorem 3.1.5 the bounded solutions of equation (3.1.1) need not oscillate.

Example 3.1.6   The neutral delay differential equation

$$\frac{d^2}{dt^2}\left[y(t) - \frac{2\,e^{\pi/2} + e^{-\pi/2}}{e^{3\pi/2} + 2\,e^{-3\pi/2}}\, y(t - 2\pi) \right] + \frac{2(e^{2\pi} - e^{-2\pi})}{e^{3\pi/2} + 2\,e^{-3\pi/2}}\, y\left(t - \frac{\pi}{2} \right) = 0,$$

$$t \geq 0$$

satisfies the conditions of Theorem 3.1.5. Therefore, each unbounded solution of this equation oscillates. For instance, $y(t) = e^t \sin t$ is such a solution. On the other hand, the bounded solutions of this equation do not have to oscillate. For example, $y(t) = e^{-t}$ is such a solution.

Theorem 3.1.6

Consider the neutral delay differential equation (3.1.1). Assume the following conditions eventually fulfilled:

$$0 < P(t) \equiv p \quad \text{is constant,}$$

$$Q(t) \geq 0, \quad Q(t) \not\equiv 0 \quad \text{and } \tau\text{-periodic.}$$

Then every solution of equation (3.1.1) oscillates.

Proof

Assume, for the sake of contradiction, that there is an eventually positive solution $y(t)$ of equation (3.1.1). Set

$$z(t) = y(t) + py(t - \tau).$$

Then

$$z(t) > 0 \qquad (3.1.25)$$

and

$$z''(t) = -Q(t)y(t - \sigma) \leq 0.$$

Thus, $z'(t)$ is a decreasing function of t. We claim that

$$z'(t) \geq 0.$$

Otherwise,

$$z'(t) < 0 \quad \text{and} \quad z''(t) \leq 0$$

which imply that

$$\lim_{t \to \infty} z(t) = -\infty.$$

But this contradicts (3.1.25).

Now set

$$w(t) = z(t) + pz(t - \tau)$$

which is positive. Then a direct substitution shows that $w(t)$ is a continuously differentiable solution of the neutral delay differential equation

$$w''(t) + pw''(t - \tau) + Q(t)w(t - \sigma) = 0. \qquad (3.1.26)$$

Also, we have

$$w'(t) > 0$$

and

$$w''(t) = -Q(t)z(t - \sigma) \leq 0.$$

As $z(t)$ is an increasing function, it follows that

$$w''(t - \tau) = -Q(t)z(t - \sigma - \tau) \geq -Q(t)z(t - \sigma) = w''(t).$$

Hence, from equation (3.1.26), we find

$$w''(t) + \frac{Q(t)}{1+p} \, w(t-\sigma) \leqq 0.$$

Integrating this inequality from t_1 to t, with t_1 large enough, we obtain

$$w'(t) - w'(t_1) + \frac{1}{1+p} \, w(t_1-\sigma) \int_{t_1}^{t} Q(s) \, ds \leqq 0$$

which, as $t \to \infty$, leads to a contradiction. The proof of the theorem is complete.

3.2   Second order differential equations with constant coefficients

In this section a necessary and sufficient condition is obtained for oscillation of the solutions of neutral differential equations of the form

$$\frac{d^2}{dt^2} [y(t) + py(t-\tau)] + qy(t-\sigma) = 0, \quad t \geqq t_0. \tag{3.2.1}$$

Assume the following conditions (H3.2) fulfilled:
H3.2.1 $p, q \in \mathbb{R}$,
H3.2.2 $\tau, \sigma = \text{const.} > 0$.

Theorem 3.2.1

Let conditions H3.2.1–H3.2.2 hold. Then a necessary and sufficient condition for each solution of equation (3.2.1) to oscillate is that the characteristic equation

$$\lambda^2 + \lambda^2 p \, e^{-\lambda\tau} + q \, e^{-\lambda\sigma} = 0 \tag{3.2.2}$$

of equation (3.2.1) should have no real roots.

As we see from the above theorem, the oscillatory nature of the neutral delay differential equation (3.2.1) is determined by its characteristic equation (3.2.2). This result is in contrast with the fact that the stability nature of equation (3.2.1) is not determined (as in the case of nonneutral equations) by the characteristic roots of equation (3.2.2). In fact, it was shown by Snow [145] (see also Slemrod and Infante [144]) that it is possible for a neutral delay differential equation to have all of its characteristic roots in the negative half-plane Re $\lambda < 0$ and still to have unbounded solutions.

In the sequel all functional inequalities that we write are assumed to hold eventually, that is for all t large enough.

Proof of Theorem 3.2.1

Necessity. If each solution of equation (3.2.1) oscillates, then equation (3.2.2) cannot have any real root. Indeed, if λ_0 were a real root of equation (3.2.2), then

$$y(t) = e^{\lambda_0 t}$$

would be a nonoscillating solution of equation (3.2.1).

Sufficiency. Its proof is quite involved and will be accomplished by a series of claims.

When $\tau = 0$ and $p = -1$, the result is obvious. When $p\tau = 0$ and $p \neq -1$, the neutral delay differential equation reduces to a delay equation of the form

$$y''(t) + Qy(t - \sigma) = 0 \qquad (3.2.3)$$

and equation (3.2.2) to

$$G(\lambda) \equiv \lambda^2 + Q\,e^{-\lambda\sigma} = 0. \qquad (3.2.4)$$

The hypothesis that equation (3.2.4) has no real root and the fact that

$$G(\infty) = \infty$$

imply that

$$G(\lambda) > 0 \quad \text{for } \lambda \in \mathbb{R}.$$

In particular

$$G(0) = Q > 0$$

and so, by a result of Waltman [151], each solution of equation (3.2.3) oscillates.

In the sequel we shall assume, without stating it explicitly, that

$$p\tau \neq 0.$$

Claim 1.   Assume that equation (3.2.1) has no real roots. Then the following statements are true.

(i) $q > 0$.

(ii) If $p < 0$, then $\sigma > \tau$.

(iii) There exists a positive constant m such that

$$F(\lambda) \equiv \lambda^2 + \lambda^2 p e^{-\lambda\tau} + q e^{-\lambda\sigma} \geq m \quad \text{for all } \lambda \in \mathbb{R}. \qquad (3.2.5)$$

Indeed, $F(\infty) = \infty$ and since $F(\lambda) = 0$ has no real zeros,

$$F(\lambda) > 0 \quad \text{for all } \lambda \in \mathbb{R}. \qquad (3.2.6)$$

In particular,

$$F(0) = q > 0$$

which proves (i). Also (ii) should hold, for otherwise $F(-\infty) = -\infty$ and $F(\lambda) = 0$ would have a real zero. Finally, for the proof of (iii) observe that $F(-\infty) = F(\infty) = \infty$ which together with (3.2.6) implies

$$m = \min_{\lambda \in \mathbb{R}} F(\lambda)$$

is a positive number which satisfies (3.2.5).

In the sequel we assume that equation (3.2.1) has no real roots but, for the sake of contradiction, we also assume that equation (3.2.1) has an eventually positive solution $y(t)$. Set

$$v(t) = -[y(t) + p y(t - \tau)]. \qquad (3.2.7)$$

Claim 2.   (i) $v(t)$ is a twice continuously differentiable solution of equation (3.2.1). That is,

$$v''(t) + p v''(t - \tau) + q v(t - \sigma) = 0. \qquad (3.2.8)$$

(ii) One of the following holds: either

$$v(t) > 0, \quad v'(t) < 0, \quad v''(t) > 0 \quad \text{and} \quad \lim_{t \to \infty} v(t) = \lim_{t \to \infty} v'(t) = 0$$

$$(3.2.9)$$

or

$$v(t) > 0, \quad v'(t) > 0, \quad v''(t) > 0 \quad \text{and} \quad \lim_{t \to \infty} v(t) = \lim_{t \to \infty} v'(t) = \infty.$$

(3.2.10)

(iii) $p < 0$ and $\sigma > \tau$. And if (3.2.10) holds, then $p < -1$.

The proof of (3.2.8) follows from (3.2.7) and the linearity and autonomous character of equation (3.2.1). (It also follows by a direct substitution of (3.2.7) into (3.2.8). For the proof of (ii) observe that

$$v''(t) = qy(t - \sigma) > 0 \qquad (3.2.11)$$

which implies that $v'(t)$ is strictly increasing and so either

$$\lim_{t \to \infty} v'(t) = 0 \qquad (3.2.12)$$

or

$$\lim_{t \to \infty} v'(t) \equiv L \in \mathbb{R}. \qquad (3.2.13)$$

Clearly, (3.2.12) implies (3.2.10). So assume that (3.2.13) holds. First we shall prove that $L = 0$. Indeed, integrating (3.2.11) from t_0 to t and letting $t \to \infty$, we see that

$$L - v'(t_0) = q \int_{t_0}^{\infty} y(s - \sigma) \, ds$$

which shows that

$$y \in L_1[t_0, \infty).$$

Hence,

$$v \in L_1[t_0, \infty) \qquad (3.2.14)$$

and so $L = 0$. Thus $v'(t)$ increases to zero which implies that eventually

$$v'(t) < 0.$$

But then $v(t)$ decreases and in view of (3.2.14)

$$\lim_{t \to \infty} v(t) = 0.$$

Therefore, $v(t)$ decreases to zero which implies that eventually

$$v(t) > 0.$$

The proof of (ii) is complete. It is now clear from (3.2.7) and the fact that $v(t) > 0$ (in both (3.2.9) and (3.2.10)) that $p < 0$, so in view of Claim 1 (ii) we have

$$p < 0 \quad \text{and} \quad \sigma > \tau. \tag{3.2.15}$$

Finally, assume that (3.2.10) holds and that, for the sake of contradiction, $-1 \le p < 0$. Clearly, (3.2.10) implies that $y(t)$ is unbounded. Thus, there exists a sequence of points $\{t_n\}$ such that

$$\lim_{n \to \infty} t_n = \infty \quad \text{and} \quad y(t_n) = \max_{s \le t_n} y(s) \quad \text{for } n = 1, 2, \ldots.$$

Now, from (3.2.7), we see that

$$v(t_n) = -y(t_n) - py(t_n - \tau) \le -y(t_n) - py(t_n)$$
$$= -(1 + p)y(t_n) \le 0$$

which contradicts (3.2.10) and completes the proof of Claim 2.

Next, we shall define two sets corresponding to whether (3.2.9) or (3.2.10) is satisfied. Let W^- and W^+ be the sets of all functions of the form

$$w(t) = -[v(t) + pv(t - \tau)]$$

where $v(t)$ is a twice continuously differentiable solution of equation (3.2.1) which satisfies (3.2.9) and (3.2.10), respectively. In view of Claim 2, either W^- or W^+ is nonempty. Also, an argument similar to that of Claim 2 shows that each function $w \in W^- \cup W^+$ is a four times continuously differentiable solution of equation (3.2.1), that is $w \in C^4$ and

$$w''(t) + pw''(t - \tau) + qw(t - \sigma) = 0. \tag{3.2.16}$$

Also, there is a solution $v \in C^2$ of equation (3.2.1) which satisfies (3.2.9) if $w \in W^-$ or (3.2.10) if $w \in W^+$ such that

$$w''(t) = qv(t - \sigma). \tag{3.2.17}$$

Clearly, every function $w \in W^-$ satisfies

$$w(t) > 0, \quad w'(t) < 0, \quad w''(t) > 0 \quad \text{and} \quad \lim_{t \to \infty} w(t) = \lim_{t \to \infty} w'(t) = 0$$

$$(3.2.18)$$

while every function $w \in W^+$ satisfies

$$w(t) > 0, \quad w'(t) > 0, \quad w''(t) > 0 \quad \text{and} \quad \lim_{t \to \infty} w(t) = \lim_{t \to \infty} w'(t) = \infty.$$

$$(3.2.19)$$

Furthermore,

$$w(t) \in W^- \Rightarrow -[w(t) + pw(t - \tau)] \in W^-$$

and

$$w(t) \in W^+ \Rightarrow -[w(t) + pw(t - \tau)] \in W^+.$$

Finally, w_1 and $w_2 \in W^-$ (respectively in W^+) and $a, b > 0 \Rightarrow aw_1 + bw_2 \in W^-$ (respectively in W^+).

For any function $w \in W^- \cup W^+$ define the set

$$\Lambda(w) = \{\lambda \geq 0 : w''(t) - \lambda^2 w(t) \geq 0\}.$$

Clearly, $0 \in \Lambda(w)$ and if $\lambda \in \Lambda(w)$, then $[0, \lambda] \subset \Lambda(w)$. That is, $\Lambda(w)$ is a nonempty subinterval of $\mathbb{R}^+$.

First, we shall assume that $W^- \neq \emptyset$ and we shall show that this leads to a contradiction.

Claim 3.   (i) Let $w \in W^-$. Then $(q/(-p))^{1/2} \in \Lambda(w)$.

(ii) There is a constant μ (*independent of w*) such that $\Lambda(w)$ is bounded above by μ for all $w \in W^-$.

(iii) Let $w \in W^-$ and $\lambda \in \Lambda(w)$. Then $w'(t) + \lambda w(t) \leq 0$.

The proof of (i) follows from the observation that $w''(t) > 0$ and so (3.2.16) yields

$$pw''(t - \tau) + qw(t - \sigma) < 0$$

or

$$w''(t) + \frac{q}{p} w(t - (\sigma - \tau)) > 0. \qquad (3.2.20)$$

In view of (3.2.15) and the decreasing nature of $w(t)$ we have

$$w''(t) + \frac{q}{p} w(t) > 0$$

which proves (i). Next, we turn to (ii). By integrating (3.2.20) from $t - \alpha$ to t, with $\alpha > 0$, we find

$$w'(t) - w'(t - \alpha) + \frac{q}{p} \alpha w(t - (\sigma - \tau)) > 0$$

and so

$$w'(t) + \frac{q}{-p} \alpha w(t + \alpha - (\sigma - \tau)) < 0. \qquad (3.2.21)$$

Again, by integrating (3.2.21) from $t - \beta$ to t, with $\beta > 0$, we get

$$w(t) - w(t - \beta) + \frac{q}{-p} \alpha \beta w(t + \alpha - (\sigma - \tau)) < 0$$

and so

$$\frac{q}{-p} \alpha \beta w(t + \alpha + \beta - (\sigma - \tau)) < w(t).$$

Taking $\alpha = \beta - \frac{1}{4}(\sigma - \tau)$, we see that

$$w\left(t - \frac{\sigma - \tau}{2} \right) < A w(t) \qquad (3.2.22)$$

where

$$A = \frac{-16p}{q(\sigma - \tau)^2}.$$

Let k be a positive integer such that

$$\frac{\sigma - \tau}{2}(k - 1) \leq \sigma \leq \frac{\sigma - \tau}{2} k.$$

Then, from (3.2.22) and the decreasing nature of $w(t)$, we see that

$$w(t - \sigma) < A^k w(t) \qquad (3.2.23)$$

Next, integrating (3.2.17) from $t - \alpha$ to t, with $\alpha > 0$, we find

$$w'(t) - w'(t - \alpha) > q\alpha v(t - \sigma)$$

and so

$$w'(t) + q\alpha v(t + \alpha - \sigma) < 0.$$

By integrating again from $t - \beta$ to t, we find

$$w(t) - w(t - \beta) + q\alpha\beta v(t + \alpha - \sigma) < 0$$

and so

$$q\alpha\beta v(t + \alpha + \beta - \sigma) < w(t).$$

In particular, for $\alpha = \beta = \sigma/2$ we get

$$\frac{q\sigma^2}{4} v(t) < w(t). \qquad (3.2.24)$$

Therefore, from (3.2.17), (3.2.24) and (3.2.23) we obtain

$$w''(t) = qv(t - \sigma) < \frac{4}{\sigma^2} w(t - \sigma) < \frac{4A^k}{\sigma^2} w(t)$$

which shows that

$$\mu \equiv \frac{2A^{k/2}}{\sigma} \notin \Lambda(w). \qquad (3.2.25)$$

Clearly μ depends on σ, τ, p and q only and not on any particular $w \in W^-$. The proof of (ii) is complete. For the proof of (iii), set

$$\psi(t) = w'(t) + \lambda w(t).$$

Then

$$\psi'(t) - \lambda\psi(t) = w''(t) - \lambda^2 w(t) \geqq 0$$

and so

$$\frac{d}{dt}\left[\psi(t)e^{-\lambda t}\right] \geq 0.$$

This shows that $\psi(t)e^{-\lambda t}$ is an increasing function. But

$$\lim_{t \to \infty}\left[\psi(t)e^{-\lambda t}\right] = 0$$

which implies $\psi(t)e^{-\lambda t} \leq 0$ and so $\psi(t) \leq 0$. The proof of Claim 3 is complete.

For any function $w \in W^-$ by integrating (3.2.16) from t to t_1 twice and by letting $t_1 \to \infty$ and using (3.2.18) we find that

$$-[w(t) + pw(t-\tau)] = q \int_t^\infty \int_s^\infty w(\eta - \sigma)\, d\eta\, ds. \qquad (3.2.26)$$

In particular, (3.2.26) shows that the right-hand side of (3.2.26) is an element of W^-. Also, for any $w \in W^-$ and any $\lambda \in \mathbb{R}$ the function

$$z(t) = -[w(t) + pw(t-\tau)] + \lambda^2 \int_{t-\sigma}^\infty \int_s^\infty w(\eta)\, d\eta\, ds \qquad (3.2.27)$$

is an element of W^-.

Claim 4.   Let $w \in W^-$ and let $\lambda \in \Lambda(w)$. Set

$$k = \frac{m}{-p\,e^{\mu\tau} + e^{\mu\sigma}}$$

where m is the constant which was defined in Claim 1 (iii) and μ is the constant which was defined in Claim 3 (ii). Then

$$(\lambda^2 + k)^{1/2} \in \Lambda(z),$$

where z is the function defined by (3.2.27).
    Indeed, from (3.2.27) we see that

$$z''(t) = (q + \lambda^2)w(t-\sigma).$$

As $\lambda \in \Lambda(w)$,

$$-w''(t) + \lambda^2 w(t) \leq 0$$

and so

$$-w(t) + \lambda^2 \int_t^\infty \int_s^\infty w(\eta) \, d\eta \, ds \leq -w(t) + \int_t^\infty \int_s^\infty w''(\eta) \, d\eta \, ds$$

$$= -w(t) + w(t) = 0.$$

Hence

$$z(t) \leq -pw(t-\tau) + \lambda^2 \int_{t-\sigma}^t \int_s^\infty w(\eta) \, d\eta \, ds. \qquad (3.2.29)$$

Using (3.2.28) and (3.2.29), we have

$$-z''(t) + (\lambda^2 + k)z(t) \leq -(q + \lambda^2)w(t - \sigma) - (\lambda^2 + k)pw(t - \tau)$$

$$+ (\lambda^2 + k)\lambda^2 \int_{t-\sigma}^t \int_s^\infty w(\eta) \, d\eta \, ds. \qquad (3.2.30)$$

Set

$$q(t) = e^{\lambda t} w(t).$$

Then, by Claim 3 (iii),

$$\varphi'(t) = e^{\lambda t}[w'(t) + \lambda w(t)] \leq 0$$

which shows that $\varphi(t)$ is a decreasing function. Observe that

$$\lambda^2 \int_{t-\sigma}^t \int_s^\infty w(\eta) \, d\eta \, ds = \lambda^2 \int_{t-\sigma}^t \int_s^\infty e^{-\lambda \eta} \varphi(\eta) \, d\eta \, ds$$

$$\leq \lambda^2 \varphi(t - \sigma) \int_{t-\sigma}^t \int_s^\infty e^{-\lambda \eta} \, d\eta \, ds$$

$$= \varphi(t - \sigma) e^{-\lambda t}(e^{\lambda \sigma} - 1).$$

Using this inequality, the fact that $\sigma > \tau$, the decreasing nature of $\varphi(t)$, and (3.2.5), (3.2.30) yields

$$-z''(t) + (\lambda^2 + k)z(t)$$
$$\leq e^{-\lambda t}\,\varphi(t - \sigma)[-(q + \lambda^2)\,e^{\lambda\sigma} - (\lambda^2 + k)p\,e^{\lambda\tau} + (\lambda^2 + k)(e^{\lambda\sigma} - 1)]$$
$$\leq e^{-\lambda t}\,\varphi(t - \sigma)[(-\lambda^2 - \lambda^2 p\,e^{\lambda\tau} - q\,e^{\lambda\sigma}) + k(-p\,e^{\lambda\tau} + e^{\lambda\sigma})]$$
$$\leq e^{-\lambda t}\,\varphi(t - \sigma)(-m + m) = 0$$

which proves Claim 4.

Finally, consider the sequence of functions

$$z_n(t) = -[z_{n-1}(t) + pz_{n-1}(t - \tau)] + \lambda_n^2 \int_{t-\sigma}^{\infty} \int_{t}^{\infty} z_{n-1}(\eta)\,d\eta\,ds$$

for $n = 1, 2, \ldots$, where $z_0(t)$ is the function $z(t)$ defined in (3.2.27), λ_0 is the number (see Claim 3 (i))

$$\lambda_0 = (q/(-p))^{1/2}$$

and

$$\lambda_n = (\lambda_{n-1}^2 + k)^{1/2}.$$

A repeated application of Claim 4 shows that

$$\lambda_n \in \Lambda(z_{n-1}) \quad \text{for } n = 1, 2, \ldots.$$

Clearly,

$$\lim_{n \to \infty} \lambda_n = \infty$$

which contradicts the fact proved in Claim 3 (ii) that $\lambda_n \leq \mu$ for all $n = 1, 2, \ldots$. The proof of the theorem is complete when $W^- \neq \varnothing$.

Next, we shall assume that $W^- = \varnothing$. Then $W^+ \neq \varnothing$. In view of Claim 2 (iii) this implies that

$$p < -1. \tag{3.2.31}$$

The next claim is dual to Claim 3 for the case where $W^+ \neq \varnothing$.

Claim 5.   (i) Let $w \in W^+$ and $\lambda \in \Lambda(w)$. Then $w'(t) - \lambda w(t) \geq 0$.
(ii) Let k be a positive integer such that $k\tau > \sigma - \tau$. Then

$$\left[\frac{q}{(-p)^{k+1}} \right]^{1/2} \in \Lambda(w) \quad \text{for every } w \in W^+.$$

(iii)

$$\frac{1}{\tau} \ln(-p) \notin \Lambda(w) \quad \text{for any } w \in W^+.$$

For the proof of (i) set $\theta(t) = e^{-\lambda t} w(t)$ and observe that

$$\theta'(t) = e^{-\lambda t}[w'(t) - \lambda w(t)]$$

$$\theta''(t) = e^{-\lambda t}[w''(t) - 2\lambda w'(t) + \lambda^2 w(t)]$$

and

$$\theta''(t) + 2\lambda \theta'(t) = e^{-\lambda t}[w''(t) - \lambda^2 w(t)] \geq 0. \qquad (3.2.32)$$

From (3.2.32) we see that $\theta(t)\,e^{2\lambda t}$ is a nondecreasing function and so if the claim were false, then

$$\theta'(t) < 0. \qquad (3.2.33)$$

From (3.2.33) and (3.2.32) we see that

$$\theta''(t) > 0$$

and so

$$w''(t) - 2\lambda w'(t) + \lambda^2 w(t) > 0$$

which together with the hypothesis that

$$w''(t) - \lambda^2 w(t) \geq 0$$

implies that

$$w''(t) - \lambda w'(t) > 0. \qquad (3.2.34)$$

Set

$$u(t) = -w'(t) + \lambda w(t).$$

Then, $u(t)$ is a solution of equation (3.2.1) and because of (3.2.33) and (3.2.34)

$$u(t) > 0 \quad \text{and} \quad u'(t) < 0. \tag{3.2.35}$$

Now using u instead of y in (3.2.7) and the hypothesis that $W^- = \emptyset$ we see, as in the proof of (3.2.10), that

$$\lim_{t \to \infty} [-(u(t) + pu(t - \tau))] = \infty.$$

But (3.2.35) implies that

$$\lim_{t \to \infty} u(t) \in \mathbb{R}$$

and this contradiction completes the proof of (i).

For the proof of (ii), let $w \in W^+$. Then

$$-[w(t) + pw(t - \tau)] \in W^+$$

and so

$$-pw(t - \tau) > w(t). \tag{3.2.36}$$

It follows by iteration and the increasing nature of $w(t)$ that

$$w(t) < (-p)^k w(t - k\tau) < (-p)^k w(t - (\sigma - \tau)).$$

From (3.2.16) we see that

$$pw''(t - \tau) + qw(t - \sigma) < 0$$

or

$$0 < w''(t) - \frac{q}{-p} w(t - (\sigma - \tau)) < w''(t) - \frac{q}{(-p)^{k+1}} w(t)$$

and the proof of (ii) is complete.

For the proof of (iii) assume, for the sake of contradiction, that $\lambda_0 \equiv (1/\tau)\ln(-p)\in\Lambda(w)$ for some $w\in W^+$. Then

$$w''(t) - \lambda_0^2 w(t) \geq 0$$

and by (i)

$$w'(t) - \lambda_0 w(t) \geq 0.$$

Thus the function

$$w(t)\,e^{-\lambda_0 t}$$

is increasing which implies that

$$w(t-\tau)\,e^{-\lambda_0(t-\tau)} \leq w(t)\,e^{-\lambda_0 t}$$

or

$$w(t-\tau)\,e^{\lambda_0\tau} \leq w(t).$$

But from the definition of λ_0,

$$e^{\lambda_0\tau} = -p$$

and so

$$-pw(t-\tau) \leq w(t)$$

which contradicts (3.2.36). The proof of Claim 5 is complete.

Integrating both sides of (3.2.16) from $t_0+\sigma$ to $t-\tau$ we find

$$[w'(t-\tau) + pw'(t-2\tau)]$$

$$-[w'(t_0+\sigma) + pw'(t_0+\sigma-\tau)] + q\int_{t_0+\sigma}^{t-\tau} w(s-\sigma)\,ds = 0$$

or

$$-[w'(t+\sigma-\tau) + pw'(t+\sigma-2\tau)] = c + q\int_{t_0}^{t-\tau} w(s)\,ds \quad (3.2.37)$$

where

$$c = -[w'(t_0 + \sigma) + pw'(t_0 + \sigma - \tau)]. \qquad (3.2.38)$$

As $w'(t)$ is a solution of equation (3.2.1), it follows from (3.2.37) that if $w \in W^+$, then

$$c + q \int_{t_0}^{t-\tau} w(s) \, ds \in W^+$$

where c is the constant given by (3.2.38).

Claim 6. Let $w \in W^+$ and let $\lambda \in \Lambda(w)$, $\lambda \geq \lambda_0 \equiv \{q/[(-p)^{k+1}]\}^{1/2}$. Set

$$N = \frac{m}{2(-p + q\lambda_0^{-2})}$$

where m is the constant which was defined in Claim 1 (iii). Then

$$(\lambda^2 + N)^{1/2} \in \Lambda(z)$$

where

$$z(t) = -[w(t) + pw(t - \tau)] + \frac{q}{\lambda} \int_{t_0}^{t-\tau} w(s) \, ds + \frac{c}{\lambda}$$

and c is the constant given by (3.2.38).

Clearly, $z \in W^+$. We have

$$z''(t) - (\lambda^2 + N)z(t) = qw(t - \sigma) + \frac{q}{\lambda} w'(t - \tau) + \lambda^2 w(t)$$

$$+ \lambda^2 pw(t - \tau) - q\lambda \int_{t_0}^{t-\tau} w(s) \, ds + Nw(t)$$

$$+ Npw(t - \tau) - \frac{qN}{\lambda} \int_{t_0}^{t-\tau} w(s) \, ds + c_1 \quad (3.2.39)$$

where

$$c_1 = -(\lambda^2 + N)\frac{c}{\lambda}.$$

From Claim 5 (i) we have

$$w'(t) \geq \lambda w(t) \tag{3.2.40}$$

and so integrating from t_0 to $t - \sigma$, for t_0 large enough, we have

$$w(t - \sigma) - w(t_0) \geq \lambda \int_{t_0}^{t-\sigma} w(s)\, ds$$

or

$$w(t - \sigma) \geq \lambda \int_{t_0}^{t-\sigma} w(s)\, ds. \tag{3.2.41}$$

Set

$$\varphi(t) = e^{-\lambda t} w(t).$$

Then

$$\varphi'(t) = e^{-\lambda t}[w'(t) - \lambda w(t)] \geq 0$$

which shows that $\varphi(t)$ is an increasing function of t. Hence

$$w(t - \tau)\, e^{\lambda \tau} \leq w(t) \tag{3.2.42}$$

Using (3.2.40), (3.2.41) and (3.2.42) in (3.2.39), we find that

$$z''(t) - (\lambda^2 + N)z(t) \geq q\lambda \left[\int_{t_0}^{t-\sigma} w(s)\, ds - \int_{t_0}^{t-\tau} w(s)\, ds \right] + qw(t - \tau)$$

$$+ \lambda^2\, e^{\lambda t}\, w(t - \tau) + \lambda^2 pw(t - \tau) + Nw(t)$$

$$+ Npw(t - \tau) - \frac{qN}{\lambda^2}[w(t - \tau) + w(t_0)] + c_1$$

$$\geq -q\lambda \int_{t-\sigma}^{t-\tau} w(s)\, ds + qw(t - \tau) + \lambda^2\, e^{\lambda \tau} w(t - \tau)$$

$$+ \lambda^2 pw(t - \tau) + Npw(t - \tau) - \frac{qN}{\lambda^2} w(t - \tau) + c_1.$$

$$\tag{3.2.43}$$

Also $\varphi(t)$ is an increasing function of t which implies that

$$
-\lambda \int_{t-\sigma}^{t-\tau} w(s)\,ds = -\lambda \int_{t-\sigma}^{t-\tau} e^{\lambda s}\,\varphi(s)\,ds
$$

$$
\geq \varphi(t-\tau)\left[-\lambda \int_{t-\sigma}^{t-\tau} e^{\lambda s}\,ds\right]
$$

$$
= -\varphi(t-\tau)[e^{\lambda(t-\tau)} - e^{\lambda(t-\sigma)}]
$$

$$
= -e^{\lambda(t-\tau)}\,\varphi(t-\tau)[1 - e^{-\lambda(\sigma-\tau)}]
$$

$$
= -w(t-\tau)[1 - e^{-\lambda(\sigma-\tau)}]. \tag{3.2.44}
$$

Using (3.2.44) in (3.2.43), we find

$$
z''(t) - (\lambda^2 + N)z(t) \geq w(t-\tau)\left[-q + qe^{-\lambda(\sigma-\tau)} + q + \lambda^2 e^{\lambda\tau} + \lambda^2 p \right.
$$

$$
\left. - N\left(-p + \frac{q}{\lambda_0^2}\right) + \frac{c_1}{w(t-\tau)}\right]
$$

$$
= w(t-\tau)e^{\lambda\tau}\left[qe^{-\lambda\sigma} + \lambda^2 + \lambda^2 p e^{-\lambda\tau}\right.
$$

$$
\left. - N\left(-p + \frac{q}{\lambda_0^2}\right)e^{-\lambda\tau} + \frac{c_1}{w(t-\tau)}e^{-\lambda\tau}\right]
$$

$$
\geq w(t-\tau)e^{\lambda\tau}\left[m + \frac{c_1}{w(t-\tau)}e^{-\lambda\tau} - N\left(-p + \frac{q}{\lambda_0^2}\right)e^{-\lambda\tau}\right].
$$

As $\lim_{t\to\infty} w(t) = \infty$, it follows that for t large enough,

$$
m + \frac{c_1}{w(t-\tau)}e^{-\lambda\tau} \geq \frac{m}{2}.
$$

Thus

$$
z''(t) - (\lambda^2 + N)z(t) \geq w(t-\tau)e^{\lambda\tau}\left[\frac{m}{2} - N\left(-p + \frac{q}{\lambda_0^2}\right)\right] = 0
$$

which completes the proof of Claim 5 that

$$
(\lambda^2 + N)^{1/2} \in \Lambda(z).
$$

Finally, consider the sequence of functions

$$z_n(t) = -[z_{n-1}(t) + pz_{n-1}(t-\tau)] + \frac{q}{\lambda_n} \int_{t_0}^{t-\tau} z_{n-1}(s)\,ds + \frac{c_n}{\lambda_n}$$

for $n = 1, 2, \ldots$, where $z_0(t)$ is the function $z(t)$ of Claim 6; λ_0 is as in Claim 6,

$$c_n = -[z_n'(t_0 + \sigma) + pz_n'(t_0 + \sigma - \tau)]$$

and

$$\lambda_n = (\lambda_{n-1}^2 + N)^{1/2}$$

where N is the positive constant defined in Claim 6. A repeated application of Claim 6 shows that

$$\lambda_n \in \Lambda(z_{n-1}) \quad \text{for } n=1, 2, \ldots.$$

Clearly,

$$\lim_{n \to \infty} \lambda_n = \infty$$

which contradicts the fact proved in Claim 5 (iii) that

$$\frac{1}{\tau} \ln(-p)$$

is an upper bound of $\Lambda(w)$ for $w \in W^+$. The proof of the theorem is complete.

3.3   Second order differential equations with distributed delay

In this section some oscillatory and asymptotic properties of the solutions of equations of the form

$$\frac{d^2}{dt^2}\left[x(t) + \int_0^{\sigma(t)} x(t-s)\,d_s r(t,s)\right] + \int_0^{\sigma(t)} x(t-s)\,d_s \bar{r}(t,s) = 0 \qquad (3.3.1)$$

are investigated, where $t \geq t_0$, $t_0 \in \mathbb{R}$.

We shall say that conditions (H3.3) are met if the following conditions hold:

H3.3.1   The functions r, $\bar{r}:[t_0, \infty) \times [0, \infty) \to \mathbb{R}$ are measurable as functions of two variables, nondecreasing with respect to the second argument, $r(t,0) \equiv \bar{r}(t,0) \equiv 0$ and the following relations are valid (see [123])

$$\lim_{t' \to t} \int_0^{\min(\sigma(t'),\sigma(t))} |r(t', \tau) - r(t, \tau)| \, d\tau = 0,$$

$$\lim_{t' \to t} \int_0^{\min(\sigma(t'),\sigma(t))} |\bar{r}(t', \tau) - \bar{r}(t, \tau)| \, d\tau = 0.$$

H3.3.2   The functions $\sigma:[t_0, \infty) \to [0, \infty)$, $t \to v(t) = r(t, \sigma(t))$, $t \to \bar{v}(t) = \bar{r}(t, \sigma(t))$, $t \geq t_0$ are continuous.

H3.3.3 $\lim_{t \to \infty} [t - \sigma(t)] = \infty$.

Definition 3.3.1   A property $Q(t)$ is said to be eventually fulfilled if it is fulfilled for all sufficiently large values of t.

Definition 3.3.2   The function x defined for all sufficiently large values of t is said to be an eventual solution of equation (3.3.1) if for all sufficiently large values of t the functions x and y'' are continuous, where

$$y(t) = y_x(t) = x(t) + \int_0^{\sigma(t)} x(t - s) \, d_s r(t, s) \qquad (3.3.2)$$

and x eventually satisfies equation (3.3.1). This solution is said to be regular if it is not eventually identically equal to zero.

Definition 3.3.3   The regular solution x is said to be nonoscillating if eventually either $x(t) \geq 0$ or $x(t) \leq 0$, and otherwise it is said to oscillate. The set of all regular solutions for which eventually we have $x(t) \geq 0$ will be denoted by Ω^+.

Lemma 3.3.1

Let conditions ($H3.3$) hold and $x \in \Omega^+$. Then eventually $y(t) \geq 0$, $y'(t) \geq 0$, $y''(t) \leq 0$ and $y(\infty) > 0$.

The proof of Lemma 3.3.1 follows immediately from (3.3.1) and (3.3.2).

Theorem 3.3.1

Let conditions ($H3.3$) hold and let the following relation be valid

$$\int_{t_0}^{\infty} \bar{v}(t)\, dt = \infty. \tag{3.3.3}$$

Then each nonoscillating eventual solution x enjoys the property $\lim_{t \to \infty} \inf |x(t)| = 0$.

Proof

Let $x \in \Omega^+$, let relation (3.3.3) be valid and assume that $\lim_{t \to \infty} \inf x(t) = a > 0$. Then, integrating from t_1 large enough to $t_2 > t_1$, we obtain

$$y'(t_2) - y'(t_1) + \int_{t_1}^{t_2} \bar{v}(t)\, dt \leq 0.$$

From the last inequality, passing to the limit at $t_2 \to \infty$, we obtain that

$$\int_{t_1}^{\infty} \bar{v}(t)\, dt < \infty$$

which contradicts (3.3.3).

Corollary 3.3.1   *Let the conditions of Theorem 3.3.1 and let the following relation be valid*

$$\sup_{t \geq t_0} v(t) < \infty. \tag{3.3.4}$$

Then each nonoscillating eventual solution x of equation (3.3.1) enjoys the property $\lim_{t \to \infty} \sup |x(t)| > 0$.

Proof

Let $x \in \Omega^+$ and suppose that the assertion is not true. Then from Theorem 3.3.1 it follows that $\lim_{t \to \infty} x(t) = 0$ and in view of the inequality

$$y(t) \leq x(t) + \left[\max_{t - \sigma(t) \leq \tau \leq t} x(\tau) \right] v(t)$$

we conclude that $y(\infty) = 0$, which contradicts the assertion of Lemma 3.3.1.

Remark 3.3.1   It is immediately seen that if conditions (3.3.3) and (3.3.4) hold simultaneously, then no regular eventual solution of equation (3.3.1) can be eventually monotone.

Theorem 3.3.2

Let conditions (H3.3) hold, $\sup_{t \leq t_0} v(t) \leq 1$ *and let the following relation be valid*

$$\int_{t_0}^{\infty} \left(\int_0^{\sigma(t)} [1 - v(t - s)] \, d_s \bar{r}(t, s) \right) dt = \infty. \tag{3.3.5}$$

Then all regular eventual solutions of equation (3.3.1) oscillate.

Proof

Suppose that $x \in \Omega^+$. Then from equality (3.3.2) and Lemma 3.3.1 it follows that eventually the following inequality holds

$$x(t) \geq y(t) - \int_0^{\sigma(t)} y(t - s) \, d_s r(t, s) \geq [1 - v(t)] y(t).$$

In view of equation (3.3.1), from the last inequality it follows that eventually the following inequality holds

$$y''(t) + \int\limits_{0}^{\sigma(t)} [1 - v(t-s)]y(t-s)\,\mathrm{d}_s\bar{r}(t,s) \leqq 0. \qquad (3.3.6)$$

Integrate inequality (3.3.6) from t_1 large enough to $t_2 > t_1$ and in view of Lemma 3.3.1 we obtain

$$\int\limits_{t_1}^{t_2}\left(\int\limits_{0}^{\sigma(t)} [1 - v(t-s)]y(t-s)\,\mathrm{d}_s\bar{r}(t,s)\right)\mathrm{d}t \leqq y'(t_1).$$

Passing to the limit in the last inequality as $t_2 \to \infty$ and taking into account that $y(t-s) \geqq \mathrm{const.} > 0$, then we obtain the relation

$$\int\limits_{t_1}^{\infty}\left(\int\limits_{0}^{\sigma(t)} [1 - v(t-s)]\,\mathrm{d}_s\bar{r}(t,s)\right)\mathrm{d}t < \infty$$

which contradicts relation (3.3.5).

Remark 3.3.2   By the change of variables $t - s \to t$ condition (3.3.5) can be given the following form. Denote for $t \in [t_0, \infty)$, $s \in [0, \infty)$ by $\tilde{r}(t,s)$ the function

$$\tilde{r}(t,s) = \begin{cases} \bar{r}(t,s), & s \in [0, \sigma(t)] \\ \bar{r}(t,\sigma(t)), & s \in (\sigma(t), \infty) \end{cases}$$

and by $\tilde{r}'_t(t,s)$ denote its derivative with respect to t in the sense of distributions. Then, setting for $t \in [t_0, \infty)$

$$\tilde{v}(t) = \int\limits_{0}^{\infty} \tilde{r}'_t(t+s,s)\,\mathrm{d}s$$

(in general, $\tilde{v}(t)$ is a distribution, more precisely, a derivative of a regular function), it is immediately seen that condition (3.3.5) is equivalent to the condition

$$\int_{t_0}^{\infty} [1 - v(t)] \tilde{v}(t) \, dt = \infty.$$

Theorem 3.3.5

Let the conditions of Theorem 3.3.1 hold and

$$\sup_{t \geq t_0} v(t) < 1$$

then all regular eventual solutions of equation (3.3.1) oscillate.

Proof

Suppose that $x \in \Omega^+$. Then as in the proof of Theorem 3.3.2 we conclude that eventually inequality (3.3.6) holds, from which in view of Lemma 3.3.1 we obtain that eventually the following inequality holds

$$y''(t) + a\bar{v}(t) \leq 0, \quad a = \text{const.} > 0.$$

Integrate the last inequality from t_1 large enough to $t_2 > t_1$ and passing to the limit as $t_2 \to \infty$, we obtain the relation

$$\int_{t_1}^{\infty} \bar{v}(t) \, dt < \infty$$

which contradicts (3.3.3).

3.4   Second order nonlinear differential equations

In this section the oscillatory properties and asymptotic behaviour of the solutions of nonlinear neutral differential equations of the form

$$\frac{d^2}{dt^2} [y(t) + P(t)y(t - \tau)] + Q(t)f(y(t - \sigma)) = 0, \quad t \geq t_0 \quad (3.4.1)$$

are investigated, where $P(t)$, $Q(t) \in C([t_0, \infty); \mathbb{R})$, $f \in C(\mathbb{R}; \mathbb{R})$ and the delays τ and σ are nonnegative constants.

We shall note that the first oscillation criterion for second order equations, valid both for linear and nonlinear neutral differential equations, was obtained by Zahariev and Bainov [160].

We shall say that conditions (H3.4) are met if the following conditions hold:

H3.4.1 $P(t)$, $Q(t) \in C([t_0, \infty); \mathbb{R})$, $f(u) \in C(\mathbb{R}; \mathbb{R})$.

H3.4.2 $Q(t) \geq 0$ for $t \in [t_0, \infty)$, $P(t) \not\equiv 0$, $Q(t) \not\equiv 0$.

H3.4.3 $uf(u) > 0$ for $u \neq 0$.

H3.4.4   If eventually the inequality

$$y(t) \geq a > 0 \quad (y(t) \leq -a < 0)$$

holds, where $a = $ const., then there exists a constant A such that eventually we have

$$f(y(t)) \geq A > 0 \quad (f(y(t)) \leq -A < 0).$$

H3.4.5 $$\int_{t_0}^{\infty} Q(s) \, ds = \infty.$$

H3.4.6   There exists a continuous function $b(t)$ such that

$$b(t) = o(t), \quad t \to \infty \quad \text{and} \quad b(t) \leq P(t) \leq 0.$$

First consider the asymptotic behaviour of the nonoscillating solutions of equation (3.4.1). In the proofs of the subsequent theorems we shall use the following lemma.

Lemma 3.4.1

Let $y(t)$ be a nonoscillating solution of (3.4.1). Then the following statements are valid for

$$z(t) = y(t) + P(t)y(t - \tau).$$

(a) Assume conditions (H3.4) fulfilled. If $y(t)$ is eventually positive, then the functions $z(t)$ and $z'(t)$ are either both decreasing with

$$\lim_{t \to \infty} z(t) = \lim_{t \to \infty} z'(t) = -\infty \tag{3.4.2}$$

or $z'(t)$ is decreasing with

$$\lim_{t \to \infty} z'(t) = 0, \quad z'(t) > 0 \quad \text{and} \quad z(t) < 0. \tag{3.4.3}$$

(b) *Assume conditions (H3.4) fulfilled. If $y(t)$ is eventually negative, then the functions $z(t)$ and $z'(t)$ are either both increasing with*

$$\lim_{t \to \infty} z(t) = \lim_{t \to \infty} z'(t) = \infty \tag{3.4.4}$$

or $z'(t)$ is increasing with

$$\lim_{t \to \infty} z'(t) = 0, \quad z'(t) < 0 \quad \text{and} \quad z(t) > 0. \tag{3.4.5}$$

(c) *Assume conditions H3.4.1–H3.4.5 fulfilled and that there exists a constant $P_1 < 0$ such that*

$$P_1 \leq P(t) \leq 0. \tag{3.4.6}$$

If $y(t)$ is eventually positive, then either (3.4.2) holds or $z'(t)$ is decreasing with

$$\lim_{t \to \infty} z(t) = \lim_{t \to \infty} z'(t) = 0, \quad z'(t) > 0 \quad \text{and} \quad z(t) < 0. \tag{3.4.7}$$

(d) *Assume conditions H3.4.1–H3.4.5 fulfilled in addition to condition (3.4.6). If $y(t)$ is eventually negative, then either (3.4.4) holds or $z'(t)$ is increasing with*

$$\lim_{t \to \infty} z(t) = \lim_{t \to \infty} z'(t) = 0, \quad z'(t) < 0 \quad \text{and} \quad z(t) > 0. \tag{3.4.8}$$

(e) *Assume conditions H3.4.1–H3.4.5 fulfilled in addition to condition (3.4.6). If $P_1 \geq -1$, then (3.4.7) holds when $y(t)$ is eventually positive and (3.4.8) holds when $y(t)$ is eventually negative.*

Proof

(a) From (3.4.1) we have that there exists $t_1 \geq t_0$ such that

$$z''(t) = -Q(t)f(y(t - \sigma)) \leq 0 \tag{3.4.9}$$

for $t \geq t_1$, so $z'(t)$ is decreasing on $[t_1, \infty)$ which implies that $z(t)$ is monotone.

If there exists $t_2 \geq t_1$ such that $z'(t_2) < 0$, then, since $Q(t) \not\equiv 0$ on any half-line, there exists $t_3 \geq t_2$ such that $z'(t) \leq z'(t_3) < 0$ for $t \geq t_3$ and so $z(t) \to -\infty$ as $t \to \infty$. Hence $z'(t) \to L \geq -\infty$. If $L = -\infty$, clearly (3.4.2) holds. If $L > -\infty$, integrate (3.4.9) to obtain

$$z'(t) = z'(t_3) - \int_{t_3}^{t} Q(s) f(y(s - \sigma)) \, ds$$

and then let $t \to \infty$ to obtain

$$\int_{t_3}^{\infty} Q(s) f(y(s - \sigma)) \, ds = z'(t_3) - L < \infty.$$

The last inequality, together with H3.4.4 and H3.4.5, implies that $\lim_{t \to \infty} \inf y(t) = 0$. Since $L < 0$, an integration shows that $z(t)$ is eventually negative, so choose $t_4 > t_3$ such that $z'(t) < L/2$ for $t \geq t_4$ and $z(t_4) < 0$. Then, by the mean value theorem for derivatives, we have

$$z(t) = z(t_4) + (t - t_4)z'(v) < L(t - t_4)(1/2)$$

for $t \geq t_4$ and hence $z(t) < \frac{1}{4}Lt$ for $t \geq 2t_4$. So from H3.4.6 we have

$$b(t)y(t - \tau) \leq P(t)y(t - \tau) < z(t) < \tfrac{1}{4}Lt.$$

Thus, we see that

$$y(t - \tau) > \frac{Lt}{4b(t)}$$

which in view of H3.4.6 implies that $y(t) \to \infty$ as $t \to \infty$ contradicting $\lim_{t \to \infty} \inf y(t) = 0$. Hence, we conclude that $z'(t) > 0$ for $t \geq t_1$ and therefore $z'(t) \to l \geq 0$ as $t \to \infty$ since $z'(t)$ is decreasing on $[t_1, \infty)$. Integrating (3.4.9) over $[t, A_1]$, $t \geq t_1$, and then letting $A_1 \to \infty$ gives

$$z'(t) = l + \int_{t}^{\infty} Q(s) f(y(s - \sigma)) \, ds$$

which again implies that $\lim_{t \to \infty} \inf y(t) = 0$. But this is impossible if $l > 0$, since $z'(t) \geq l$ and hence $y(t) \geq z(t) \to \infty$ as $t \to \infty$. Hence, we conclude that $l = 0$. Furthermore, if there exists $t_5 \geq t_1$ such that $z(t_5) \geq 0$, then $z'(t) > 0$ implies that $z(t) \geq z(t_6) > 0$ for $t \geq t_6 > t_5$, which again contradicts $\lim_{t \to \infty} \inf y(t) = 0$. Therefore, we have that $z(t) < 0$ for $t \geq t_1$ which completes the proof of (a).

    (b) The proof of (b) is similar to the proof of (a) and will be omitted.

    (c) Notice first that $(3.4.6)$ implies H3.4.6, so from (a) we have that either $(3.4.2)$ or $(3.4.3)$ holds. If $(3.4.2)$ holds, there is nothing to prove. If $(3.4.3)$ holds, then $z'(t) > 0$ and $z(t) < 0$, and from the proof of (a) we have $\lim_{t \to \infty} \inf y(t) = 0$. Thus, $z(t) \to l_1 \leq 0$. If $l_1 < 0$, then $y(t) + P_1 y(t - \tau) \leq y(t) + P(t)y(t - \tau) = z(t) < l_1$ for $t \geq t_1$. But $\lim_{t \to \infty} \inf y(t) = 0$ implies that there exists an increasing sequence $\{s_n\}$ such that $s_n \to \infty$ and $y(s_n - \tau) \to 0$ as $n \to \infty$ contradicting $y(t) > 0$. Thus, we conclude that $z(t) \to 0$ as $t \to \infty$ and the proof of (c) is complete.

    (d) The proof will be omitted since it is similar to that of (c).

    (e) Suppose that $y(t)$ is eventually positive and that $(3.4.7)$ does not hold. Then from part (c), $(3.4.2)$ holds, so that $z(t) < 0$ for all large t. Since $P_1 \geq -1$,

$$y(t) < -P(t)y(t - \tau) \leq -P_1 y(t - \tau) \leq y(t - \tau)$$

for all large t. But the last inequality implies that $y(t)$ is bounded which contradicts $(3.4.2)$. Therefore $(3.4.7)$ holds when $y(t)$ is eventually positive. The argument for the case when $y(t)$ is eventually negative is similar.

Remark 3.4.1   Parts (c) and (d) of Lemma 3.4.1 were proved in Lemma 3.1.1 for the linear case $f(u) \equiv u$ with H3.4.5 replaced by the more restrictive condition

$$Q(t) \geq q > 0 \qquad\qquad (3.4.10)$$

and $(3.4.6)$ replaced by the less restrictive condition

$$P_1 \leq P(t) \leq P_2 \qquad\qquad (3.4.11)$$

P_1 and P_2 constants. Part (e) of Lemma 3.4.1 was also proved for $f(u) \equiv u$ as a part of the lemma cited above using $(3.4.10)$ and $(3.4.11)$ with $P_1 \geq -1$.

    Having obtained results concerning the asymptotic properties of $z(t)$ and $z'(t)$, we are no ready to study the asymptotic behaviour of the nonoscillating solutions of $(3.4.1)$.

Theorem 3.4.1

Assume conditions H3.4.1–H3.4.5 fulfilled. If (3.4.6) holds with $P_1 > -1$, that is

$$-1 < P_1 \leq P(t) \leq 0 \qquad (3.4.12)$$

then each nonoscillating solution $y(t)$ of (3.4.1) satisfies $y(t) \to 0$ as $t \to \infty$.

Proof

If $y(t)$ is eventually positive, then by part (e) of Lemma 3.4.1 we have that (3.4.7) holds. Thus

$$z(t) = y(t) + P(t)y(t - \tau) < 0$$

for all t large enough. Then (3.4.12) implies that

$$y(t) < -P(t)y(t - \tau) < y(t - \tau)$$

and hence $y(t)$ is bounded.

Now suppose that $\lim_{t \to \infty} \sup y(t) = b_1 > 0$. Then there exists an increasing sequence $\{s_n\}$ satisfying $s_n \to \infty$ and $y(s_n) \to b_1$ as $n \to \infty$. Then for all large n

$$0 > z(s_n) \geq y(s_n) + P_1 y(s_n - \tau)$$

so $y(s_n - \tau) \geq y(s_n)(-P_1)^{-1}$. But this implies that $\lim_{n \to \infty} y(s_n - \tau) \geq b_1(-P_1)^{-1} > b_1$ contradicting the choice of b_1. Hence we conclude that $y(t) \to 0$ as $t \to \infty$. The proof for the case when $y(t) < 0$ is similar.

Remark 3.4.2 The conclusion of Theorem 3.4.1 was obtained in Theorem 3.1.1 for (3.4.1) when $f(u) \equiv u$ under the more restrictive hypothesis (3.4.10) and $-1 < P_1 \leq P(t) \leq P_2 < 0$. Thus Theorem 3.4.1 generalizes this result. Theorem 3.4.1 also extends special cases of Theorem 1 in [48] and Theorem 1 in [100]. Notice that some condition corresponding to H3.4.5 which prevents $Q(t)$ from being 'too small' is necessary in Theorem 3.4.1. To see this, observe that the equation

$$\frac{d^2}{dt^2}[y(t) - (t-1)^{-1/2}y(t-1)] + y^3(t-2)\frac{1}{4t^{3/2}(t-2)^{3/2}} = 0, \quad t > 2$$

has the unbounded nonoscillating solutions $y(t) = \sqrt{t}$ and satisfies all the hypotheses of Theorem 3.4.1 except for condition H3.4.5.

An example to which Theorem 3.4.1 applies is the equation

$$\frac{d^2}{dt^2}[y(t) - e^{-1}y(t-2)] + (e-1)e^{2t-3}y^3(t-1) = 0, \quad t > 0$$

which has the nonoscillating solution $y(t) = e^{-t}$.

For $P(t) \geq 0$ we have the following result.

Theorem 3.4.2

Let $P(t) \geq 0$. Then each nonoscillating solution $y(t)$ of (3.4.1) satisfies the following:

  (a) $|y(t)| \leq b_1 t$ *for some constant* $b_1 > 0$ *and all* $t \geq \max\{1, t_0\}$.
  (b) *If* $t(P(t))^{-1}$ *is bounded, then* $y(t)$ *is bounded.*
  (c) *If* $t(P(t))^{-1} \to 0$ *as* $t \to \infty$, *then* $y(t) \to 0$ *as* $t \to \infty$.

Proof

Let $y(t)$ be an eventually positive solution of (3.4.1) and let $T \geq t_0$ be such that $y(t - \tau) > 0$ and $y(t - \sigma) > 0$ for $t \geq T$. From (3.4.1) we have $z''(t) \leq 0$ for $t \geq T$, so integrating twice we have $z(t) \leq z(T) + z'(T)(t - T) \leq b_1 t$ for some constant $b_1 > 0$. Clearly $y(t) \leq b_1 t$ if $P(t) \geq 0$, so (a) holds. Moreover, $P(t)y(t - \tau) \leq b_1 t$ and hence (b) and (c) follow. The proof when $y(t)$ is eventually negative is similar.

The following equations satisfy the hypotheses of parts (a), (b) and (c) of Theorem 3.4.2, respectively, for $t \geq 4$:

$$\frac{d^2}{dt^2}[y(t) + 2y(t-1)]$$

$$+ (3t^2 - 2t + 1)y^3(t-2)[(t(t-1))^2 \ln^3(t-2)]^{-1} = 0 \qquad (3.4.13)$$

$$\frac{d^2}{dt^2}[y(t) + ty(t-1)]$$

$$+ 4(t-1)^3 y^3(t-2)(t+1)^{-3}(t-3)^{-3} = 0 \qquad (3.4.14)$$

and

$$\frac{d^2}{dt^2}[y(t) + (t-1)^2 y(t-1)] + [t^{7/2} - 15(t-1)^{3/2}]$$

$$\times (t-2)^{9/2} y^3 (t-2)[4t^{7/2}(t-1)^{3/2}]^{-1} = 0 \qquad (3.4.15)$$

The functions $y_1(t) = \ln t$, $y_2(t) = (t-1)(t+1)^{-1}$ and $y_3(t) = t^{-3/2}$ are nonoscillating solutions of (3.4.13), (3.4.14) and (3.4.15), respectively.

In the subsequent theorems results concerning the oscillatory behaviour of solutions of (3.4.1) are obtained. The first result in this direction is an immediate consequence of Lemma 3.4.1.

Theorem 3.4.3

Assume conditions H3.4.1–H3.4.5 fulfilled in addition to condition (3.4.6) with $P_1 \geqq -1$, that is,

$$-1 \leqq P(t) \leqq 0 \qquad (3.4.16)$$

Then each unbounded solution of (3.4.1) oscillates.

Proof

We need only observe that under the present hypotheses part (e) of Lemma 3.4.1 implies that all nonoscillating solutions of (3.4.1) are bounded.

Remark 3.4.3 Theorem 3.4.3 reduces to Theorem 3.1.5 in Section 3.1 and to the second order version of Theorem 12 in [48] when $f(u) \equiv u$.

In the next theorem we obtain the conclusion of Theorem 3.4.3 without requiring H3.4.5 but with more restrictive conditions on $f(u)$.

Theorem 3.4.4

Assume conditions H3.4.1–H3.4.3 fulfilled in addition to condition (3.4.16) and that f is increasing,

$$\int_t^\infty \int_s^\infty Q(v)\, dv\, ds = \infty \qquad (3.4.17)$$

and

$$\int_c^\infty \frac{1}{f(u)}\,du < \infty \quad \text{and} \quad \int_c^{-\infty} \frac{1}{f(u)}\,du < \infty \tag{3.4.18}$$

for every constant $c > 0$. Then each unbounded solution of (3.4.1) oscillates.

Proof

Suppose that (3.4.1) has an unbounded nonoscillating solution $y(t)$. If $y(t) > 0$, then $z''(t) < 0$ which implies that $z'(t)$ is decreasing and $z(t)$ is monotone. Now if $z(t)$ is eventually negative, then by (3.4.16)

$$y(t) \leq -P(t)y(t-\tau) \leq y(t-\tau)$$

contradicting the assumption that $y(t)$ is unbounded. Therefore $z(t) > 0$ eventually. Furthermore, if $z'(t)$ is eventually negative, then clearly $z(t)$ is eventually negative which, as argued above, is a contradiction. Thus, we have $z(t) > 0$ and $z'(t) > 0$ for $t \geq t_1 \geq t_0$. Since $0 < z(t) \leq y(t)$ and f is increasing, we have

$$z''(t) + Q(t)f(z(t-\sigma)) \leq 0$$

for $t \geq T = t_1 + \sigma$. For each $t \geq T$ an integration of the last inequality leads to

$$z'(t_2) - z'(t) + f(z(t-\sigma))\int_t^{t_2} Q(s)\,ds \leq 0$$

so that

$$f(z(t-\sigma))\int_t^{t_2} Q(s)\,ds \leq z'(t) - z'(t_2).$$

Letting $t_2 \to \infty$, we see that

$$f(z(t-\sigma))\int_t^\infty Q(s)\,ds \leq z'(t) \leq z'(t-\sigma)$$

for all $t \geq T$. Thus

$$\int\limits_t^\infty Q(s)\,ds \leq \frac{z'(t-\sigma)}{f(z(t-\sigma))}$$

and another integration yields

$$\int\limits_T^t \int\limits_s^\infty Q(v)\,dv\,ds \leq \int\limits_{z(T-\sigma)}^{z(t-\sigma)} \frac{1}{f(u)}\,du$$

which, in view of (3.4.18), contradicts (3.4.17).

We now give sufficient conditions for all solutions of (3.4.1) to oscillate.

Theorem 3.4.5

Assume conditions H3.4.1–H3.4.3 fulfilled, that f is increasing,

$$0 \leq P(t) \leq 1 \tag{3.4.19}$$

and

$$\int\limits_{t_0}^\infty Q(s)f([1-P(s-\sigma)]c)\,ds = \infty \tag{3.4.20}$$

for any positive constant C. Then all solutions of (3.4.1) oscillate.

Proof

Assume that (3.4.1) has a nonoscillating solution $y(t)$, say $y(t) > 0$ for $t \geq t_1 \geq t_0 + \tau + \sigma$. Notice that since $z(t) > 0$ for $t \geq t_1$, it follows by the proof of Lemma 3.4.1 (a) that $z'(t) > 0$ for $t \geq t_1$. Therefore

$$z(t-\tau) \leq z(t) \leq y(t) + P(t)z(t-\tau)$$

so

$$[1 - P(t)]z(t-\tau) \leq y(t).$$

It then follows that

$$0 = z''(t) + Q(t)f(y(t - \sigma)) \geqq z''(t) + Q(t)f([1 - P(t - \sigma)]z(t - \tau - \sigma))$$

and we see that there exists a constant $c > 0$ such that

$$z''(t) + Q(t)f([1 - P(t - \sigma)]c) \leqq 0.$$

An integration of the last inequality leads to

$$z'(t) - z'(t_1) + \int_{t_1}^{t} Q(s)f([1 - P(s - \sigma)]c)\, ds \leqq 0$$

which, since $z'(t) > 0$, contradicts (3.4.20). The proof is similar for the case when $y(t) < 0$.

Remark 3.4.4 Theorem 3.4.5 extends Theorem 1 in [49] and reduces to the second order version of Theorem 10 in [48] when $f(u) \equiv u$. When $\sigma = 0$ and $P(t) \equiv 0$, Theorem 3.4.5 reduces to a well known oscillation result for ordinary equations.

Theorem 3.4.6

Assume conditions H3.4.1–H3.4.5 fulfilled and that $P(t)$ is not eventually negative. Then any solution $y(t)$ of (3.4.1) either oscillates or satisfies $\lim_{t \to \infty} \inf |y(t)| = 0$.

Proof

Let $y(t)$ be a solution of (3.4.1). If $y(t)$ is nonoscillating, then $|y(t - \tau - \sigma)| > 0$ on $[t_1, \infty)$ for some $t_1 > t_0$. Suppose that $y(t - \tau - \sigma) > 0$ for $t \geqq t_1$. Then (3.4.9) implies that $z'(t)$ is decreasing on $[t_1, \infty)$ and hence $z(t)$ is monotone. If there exists $t_2 \geqq t_1$ such that $z'(t_2) \leqq 0$, then, since $Q(t) \not\equiv 0$ on any half-line, there exists $t_3 > t_2$ such that $z'(t) \leqq z'(t_3) < 0$ for $t \geqq t_3$ and so $z(t) \to -\infty$ as $t \to \infty$. Clearly this contradicts the hypothesis that $P(t)$ is not eventually negative. Hence $z'(t) > 0$ for $t \geqq t_1$ and therefore $z'(t) \to l \geqq 0$ as $t \to \infty$ since $z'(t)$ is decreasing. Integrating (3.4.9) over $[t, A_1]$, $t \geqq t_1$, and then letting $A_1 \to \infty$, we obtain

$$z'(t) = l + \int_t^\infty Q(s)f(y(s-\sigma))\,ds$$

which implies that $\lim_{t\to\infty} \inf y(t) = 0$. The proof when $y(t - \tau - \sigma) < 0$ for $t \geq t_1$ is similar.

Next we obtain, as a corollary to the proof of Theorem 3.4.6, a necessary condition for (3.4.1) to have a nonoscillating solution.

Corollary 3.4.1 *Suppose that (3.4.10) and (3.4.11) hold and that there exists a constant $A > 0$ such that*

$$|f(u)| \geq A|u| \quad \text{for all } u. \tag{3.4.21}$$

Then a necessary condition for (3.4.1) to have a nonoscillating solution is that $P(t)$ is eventually negative.

Proof

Notice first that (3.4.21) implies H3.4.4 and (3.4.10) implies H3.4.5. Therefore if (3.4.1) has an eventually positive solution $y(t)$, say $y(t - \tau - \sigma) > 0$ for $t \geq t_1 \geq t_0$, then from the proof of Theorem 3.4.6 we have that $z'(t)$ is decreasing and $z(t)$ is monotone on $[t_1, \infty)$. The proof of Theorem 3.4.6 also shows that $P(t) < 0$ if $z'(t)$ is eventually nonpositive, so we only need to consider the case when $z'(t) > 0$ on $[t_1, \infty)$. In this case we again have from the proof of Theorem 3.4.6 that $z'(t) \to l \geq 0$ and that

$$z'(t) = l + \int_t^\infty Q(s)f(y(s-\sigma))\,ds, \quad t \geq t_1.$$

The last equation, together with (3.4.10) and (3.4.21), implies that

$$z'(t) \geq l + Aq \int_t^\infty y(s-\sigma)\,ds$$

from which it follows that $y(t) \in L_1[t_1, \infty)$. Thus, by (3.4.11), $z(t) \in L_1[t_1, \infty)$. Since $z(t)$ is increasing, we see that $z(t) < 0$ which implies that $P(t) < 0$ in this case also. The argument when $y(t)$ is eventually negative is similar.

Corollary 3.4.1 can be restated as a sufficient condition for oscillation as follows.

Corollary 3.4.2 *If (3.4.10), (3.4.11) and (3.4.21) hold and $P(t)$ is not eventually negative, then all solutions of (3.4.1) oscillate.*

Remark 3.4.5 Corollary 3.4.2 extends Theorem 3.1 2 in Section 3.1 and the second order version of Theorem 7 in [48] and Theorem 4 in [100]. A result of Zahariev and Bainov [160], Theorem 1 includes Corollary 3.4.2 when $P(t) \equiv P > 0$ and $\tau = \sigma$. However, their method of proof does not appear to carry over under the hypotheses of Corollary 3.4.2. A similar remark can be made about the second order versions of the results in [159].

Remark 3.4.6 It seems reasonable to ask if the conclusions of Corollaries 3.4.1 and 3.4.2 can be obtained with (3.4.21) replaced by either H3.4.4 or requiring f to be increasing. Another interesting question is whether these corollaries can be proved without the requirement that $P(t)$ is not eventually negative. Theorem 3.4.3 may be considered a partial answer to the last question in case $P(t)$ is eventually negative and bounded from below by -1.

The next theorem shows that if $P(t)$ is bounded, with upper bound less than -1, then H3.4.4 and H3.4.5 are sufficient to ensure that bounded nonoscillating solutions of (3.4.1) tend to zero as $t \to \infty$.

Theorem 3.4.7

Assume conditions H3.4.1–H3.4.5 fulfilled and that there exist constants P_1 and P_3 such that

$$P_1 \leq P(t) \leq P_3 < -1. \qquad (3.4.22)$$

Then each bounded solution $y(t)$ of (3.4.1) either oscillates or satisfies $y(t) \to 0$ as $t \to \infty$.

Proof

Assume that (3.4.1) has a bounded nonoscillating solution $y(t)$. We give the details of the proof only for the case when $y(t)$ is eventually positive since the proof for $y(t)$ eventually negative is similar. By part (c) of Lemma 3.4.1 either (3.4.2) or (3.4.7) holds. Clearly (3.4.2) cannot hold in view of (3.4.22) and the fact that $y(t)$ is bounded. From (3.4.7) we have $z(t) < 0$ and $z(t) \to 0$ as $t \to \infty$. Hence for any number $\varepsilon > 0$ there exists $T \geqq t_0$ so that

$$-\varepsilon < z(t) < y(t) + P_3 y(t - \tau)$$

or

$$y(t - \tau) < (y(t) + \varepsilon)(-P_3)^{-1}.$$

So

$$y(t) < (-P_3)^{-1} y(t + \tau) + (-P_3)^{-1} \varepsilon \qquad (3.4.23)$$

and hence

$$y(t + \tau) > -P_3 y(t) - \varepsilon \qquad (3.4.24)$$

for $t \geqq T$. From (3.4.23)

$$y(t + \tau) < (-P_3)^{-1} y(t + 2\tau) + (-P_3)^{-1} \varepsilon$$

and by (3.4.24) we have

$$y(t) < (-P_3)^{-2} y(t + 2\tau) + (-P_3)^{-2} \varepsilon + (-P_3)^{-1} \varepsilon.$$

After n iterations, we obtain

$$y(t) < (-1/P_3)^n y(t + n\tau) + \sum_{i=1}^{n} (-1/P_3)^i \varepsilon.$$

Let $\lambda = 1 + (-P_3)^{-1} > 0$. Since $y(t)$ is bounded, there exists a constant $M > 0$ such that $y(t) < M$. Now choose n large enough so that $(-1/P_3)^n < \varepsilon/\lambda M$. Then we have

$$y(t) < \varepsilon \lambda^{-1} + \varepsilon [1 - (-P_3)^{-n}](1 + P_3^{-1})^{-1} < 2\varepsilon \lambda^{-1}.$$

Since ε is arbitrary, the last inequality implies that $y(t) \to 0$ as $t \to \infty$ and the proof is complete.

We conclude with an oscillation theorem for (3.4.1) when $Q(t)$ is τ-periodic.

Theorem 3.4.8

Assume conditions H3.4.1–H3.4.3 fulfilled, that $P(t) \equiv p > 0$, $Q(t)$ is τ-periodic and that f is increasing and satisfies

$$f(u + v) \leq f(u) + f(v) \quad \text{if} \quad u, v > 0,$$

$$f(u + v) \geq f(u) + f(v) \quad \text{if} \quad u, v < 0$$

$$f(ku) \leq kf(u) \quad \text{if} \quad k > 0 \quad \text{and} \quad u > 0 \qquad (3.4.25)$$

and

$$f(ku) \geq kf(u) \quad \text{if} \quad k > 0 \quad \text{and} \quad u < 0. \qquad (3.4.26)$$

Then each solution of (3.4.1) oscillates.

Proof

Let $y(t)$ be a nonoscillating solution of (3.4.1), say $y(t) > 0$ for $t \geq t_1 \geq t_0$. Then there exists $t_2 \geq t_1$ such that $z(t) > 0$ for $t \geq t_2$. From the proof of part (a) of Lemma 3.4.1, either $z'(t) > 0$ or $z(t) \to -\infty$ as $t \to \infty$. Clearly the latter cannot hold. Hence $z'(t) > 0$ for $t \geq t_3$ for some $t_3 \geq t_2$. Let $w(t) = z(t) + pz(t - \tau)$; then

$$w''(t) + pw''(t - \tau) + Q(t)f(w(t - \sigma))$$

$$= -Q(t)f(y(t - \sigma)) - 2pQ(t - \tau)f(y(t - \tau - \sigma))$$

$$\quad - p^2 Q(t - 2\tau)f(y(t - 2\tau - \sigma)) + Q(t)f(y(t - \sigma))$$

$$\quad + 2py(t - \tau - \sigma) + p^2 y(t - 2\tau - \sigma)$$

$$\leq -Q(t)f(y(t - \sigma)) - 2pQ(t)f(y(t - \tau - \sigma)) - p^2 Q(t)f(y(t - 2\tau - \sigma))$$

$$\quad + Q(t)f(y(t - \sigma)) + 2pQ(t)f(y(t - \tau - \sigma)) + p^2 Q(t)f(y(t - 2\tau - \sigma))$$

$$= 0.$$

Next observe that $w'(t) > 0$ for $t \geq t_4$ for some $t_4 \geq t_3$, so $w(t - \sigma)$ is increasing for $t \geq t_5$ for some $t_5 \geq t_4$. Integrating, we have

$$w'(t) - w'(t_5) + pw'(t - \tau) - pw'(t_5 - \tau) + \int_{t_5}^{t} Q(s)f(w(s - \sigma))\,\mathrm{d}s \leq 0.$$

It follows that

$$w'(t) - w'(t_5) + pw'(t - \tau) - pw'(t_5 - \tau) + f(w(t_5 - \sigma)) \int_{t_5}^{t} Q(s)\, ds \leqq 0.$$

Since H3.4.5 holds, we have that $w'(t) \to -\infty$ as $t \to \infty$ which is a contradiction.

Remark 3.4.7 It can easily be seen from the proof of Theorem 3.4.8 that conditions (3.4.25) and (3.4.26) need only to hold for $k = p$ and p^2. Thus it is possible to conclude that all solutions of

$$\frac{d^2}{dt^2} [y(t) + 2y(t - 2\pi)] + (1 + \sin t) y^{1/3} (t - \pi) = 0$$

oscillate.

Remark 3.4.8 Theorem 3.4.8 includes Theorem 3.1.6 in Section 3.1 and the second order version of Theorem 9 in [48] as special cases.

3.5 Distribution of the zeros of the solutions of neutral differential equations

In this section theorems estimating the distance between the zeros of the solutions of neutral differential equations of second order are obtained. We shall note that the first investigations in this direction for first order neutral differential equations were by Domshlak [22] and Domshlak and Sheikhzamanova [23].

Consider the neutral differential inequalities

$$l[x] \equiv x''(t) + \lambda(t)x''(r(t)) + c(t)x'(t) - \mu(t)x'(r(t))$$
$$+ a(t)x(t) + b(t)x(r(t)) \leqq 0, \quad t \in [t_1, t_2]. \tag{3.5.1}$$

$$\tilde{l}[y] \equiv (r'(t)y(t))'' + q'^2(t)(\lambda(q(t))y(q(t)))'' + q''(t)(\lambda(q(t))y(q(t)))'$$
$$- (c(t)r'(t)y(t))' + q'(t)(\mu(q(t))y(q(t)))'$$
$$+ \tilde{a}(t)y(t)r'(t) + \tilde{b}(q(t))y(q(t)) \geqq 0, \quad t \in [t_1, t_2]. \tag{3.5.2}$$

We shall say that conditions (H3.5) are met if the following conditions hold:

H3.5.1 $r(t) \in C^3([t_1, t_2]; \mathbb{R})$,

$$r(t) \leq t, \quad r'(t) > 0, \quad r''(t) \leq 0 \quad \text{for} \quad t \in [t_1, t_2], \quad r(t_2) > t_1.$$

H3.5.2   The function $q(t): [r(t_1), r(t_2)] \to [t_1, t_2]$ is inverse to the function $r(t)$.

H3.5.3 $a(t), \tilde{a}(t), b(t), c(t) \in C^1([t_1, t_2]; \mathbb{R})$.

H3.5.4 $\mu(t), \tilde{b}(t) \in C^1([t_1, q(t_2)]; \mathbb{R})$.

H3.5.5 $\lambda(t) \in C^2([t_1, q(t_2)]; \mathbb{R})$.

A straightforward verification shows that from conditions H3.5.1–H3.5.2 it follows that $q(t) \geq t$, $q'(t) > 0$ and $q''(t) \geq 0$ for $t \in [r(t_1), r(t_2)]$. Moreover, we assume that the function $r(t)$ is continued smoothly outside the interval $[t_1, t_2]$ so that the range of $r(t)$ should include the interval $[t_1, t_2]$. Hence we can assume that the inverse function $q(t)$ is defined in the interval $[r(t_1), t_2]$ and $q(r(t)) = r(q(t)) = t$ for $t \in [t_1, t_2]$.

Definition 3.5.1   The function $x(t)(y(t))$ is said to be a solution of the differential inequality (3.5.1) ((3.5.2)) if $x(t) \in C^2([r(t_1), t_2]; \mathbb{R})$ $(y(t) \in C^2([t_1, q(t_2)]; \mathbb{R}))$ and $x(t)$ $(y(t))$ satisfies (3.5.1) ((3.5.2)) for $t \in [t_1, t_2]$.

We shall prove the following comparison theorem of Sturmian type.

Theorem 3.5.1

Assume the following conditions fulfilled:
(1) Conditions (H3.5) hold.

(2) $\tilde{b}(t) \geq 0 \quad \text{for} \quad t \in [t_2, q(t_2)], \quad \lambda(t_i) \geq 0, \quad i = 1, 2.$ (3.5.3)

(3) Inequality (3.5.2) has a solution $y(t)$ such that:

$$y(t_1) = y(t_2) = 0, \tag{3.5.4}$$

$$y(t) > 0 \qquad \text{for} \quad t \in (t_1, t_2) \tag{3.5.5}$$

$$y(t) \leq 0 \qquad \text{for } t \in [t_2, q(t_2)] \tag{3.5.6}$$

$$(-1)^i \left[\frac{(\lambda(t)y(t))'}{r'(t)} \right]' \leq 0 \qquad \text{for } t \in [t_i, q(t_i)], \quad i = 1, 2, \tag{3.5.7}$$

$$(-1)^i (\mu(t)y(t))' \leq 0 \qquad \text{for} \quad t \in [t_i, q(t_i)], \quad i = 1, 2, \tag{3.5.8}$$

(4) $$a(t) \geq \tilde{a}(t) \quad for \quad t \in [t_1, t_2], \tag{3.5.9}$$

$$b(t) \geq \begin{cases} 0, & t \in [t_1, q(t_1)] \\ \tilde{b}(t), & t \in [q(t_1), t_2] \end{cases} \tag{3.5.10}$$

(5) *At least one of inequalities (3.5.3), (3.5.7)–(3.5.10) is fulfilled strictly at least at one point of the respective interval.*

Then inequality (3.5.1) has no solution $x(t)$ which is positive in the interval $[r(t_1), t_2]$.

Proof

Suppose, for the sake of contradiction, that there exists a solution $x(t)$ of inequality (3.5.1) which is positive in the interval $[r(t_1), t_2]$. It is immediately verified that for any solution $x(t)$ of (3.5.1) and for any smooth function $y(t)$ for which $y(t) \geq 0$ for $t \in [t_1, t_2]$ the following basic inequality holds

$$\int_{t_1}^{t_2} \tilde{l}[y] x(t)\, dt + \int_{t_1}^{q(t_1)} \left[\frac{(\lambda(t) y(t))'}{r'(t)} \right]' x(r(t))\, dt - \int_{t_2}^{q(t_2)} \left[\frac{(\lambda(t) y(t))'}{r'(t)} \right]'$$

$$x(r(t))\, dt + \int_{t_1}^{q(t_1)} [\mu(t) y(t)]' x(r(t))\, dt - \int_{t_2}^{q_2} [\mu(t) y(t)]' x(r(t))\, dt$$

$$+ \int_{t_1}^{t_2} (a(t) - \tilde{a}(t)) r'(t) y(t) x(t)\, dt + \int_{r(t_1)}^{t_1} b(q(t)) y(q(t)) x(t)\, dt$$

$$+ \int_{t_1}^{r(t_2)} (b(q(t)) - \tilde{b}(q(t))) y(q(t)) x(t)\, dt - \int_{r(t_1)}^{t_2} \tilde{b}(q(t)) y(q(t)) x(t)\, dt$$

$$\leq y'(t_2) \left[r'(t_2) x(t_2) + \frac{\lambda(t_2)}{r'(t_2)} x(r(t_2)) \right]$$

$$- y'(t_1) \left[r'(t_1) x(t_1) + \frac{\lambda(t_1)}{r'(t_1)} x(r(t_1)) \right]$$

$$+ \sum_{i=1}^{2} (-1)^{i+1} y(t_i) \left[x'(t_i) r'(t_i) + \lambda(t_i) x'(r(t_i)) - \frac{\lambda'(t_i)}{r'(t_i)} x(r(t_i)) \right.$$

$$\left. + c(t_i) r'(t_i) x(t_i) - r''(t_i) x(t_i) - \mu(t_i) x(r(t_i)) \right].$$

$$\tag{3.5.11}$$

Let $y(t)$ be a solution of (3.5.2) satisfying conditions (3.5.4)–(3.5.8). Then from conditions (3.5.3), (3.5.9), (3.5.10) it follows that all addends on the left-hand side of inequality (3.5.11) are nonnegative and at least one of them is strictly positive. On the other hand, the right-hand side of inequality (3.5.11) is nonpositive since from condition (3.5.4) it follows that $y'(t_1) \geq 0$, $y'(t_2) \leq 0$ and from H3.5.1 and condition (3.5.3) it follows that

$$\frac{\lambda(t_i)}{r'(t_i)} x(r(t_i)) + r'(t_i)x(t_i) \geq 0, \quad i = 1, 2.$$

The contradiction obtained shows that the assumption of existence of a solution $x(t)$ of inequality (3.5.1) which is positive in the interval $[r(t_1), t_2]$ is not true.

This completes the proof of Theorem 3.5.1.

Corollary 3.5.1   *Suppose that the conditions of Theorem 3.5.1 hold. Then the differential inequality $l[x] \geq 0$ has no solution $x(t)$ which is negative in the interval $[r(t_1), t_2]$.*

Corollary 3.5.2   *Suppose that the conditions of Theorem 3.5.1 hold. Then each solution of the equation*

$$l[x] = 0, \quad t \in [t_1, t_2] \tag{3.5.12}$$

has at least one zero in the interval $[r(t_1), t_2]$.

Corollary 3.5.2 can be used to obtain concrete criteria for oscillation of the solutions of equation (3.5.12) or to estimate from above the intervals between the adjacent zeros of the solutions. For this it is necessary to construct 'standard' inequalities (3.5.2) having solutions $y(t)$ with the properties (3.5.4)–(3.5.8).

Let $\varphi(t)$, $z(t)$ and $k(t)$ be functions satisfying the following conditions:
H3.5.6 $\varphi(t), z(t), k(t) \in C^1([t_1, q(t_2)]; \mathbb{R})$.

H3.5.7 $\displaystyle\int_{t_1}^{t_2} \varphi(s)\, ds = \pi.$

H3.5.8 $0 < \displaystyle\int_{t_1}^{t} \varphi(s)\, ds < \pi$ for $t \in (t_1, t_2)$.

H3.5.9 $\quad 0 < \displaystyle\int_{t_2}^{t} \varphi(s)\,ds < \pi$ for $t \in [t_2, q(t_2)]$.

H3.5.10 $\quad k(t) \le \varphi(t)\cotg \displaystyle\int_{t_i}^{t} \varphi(s)\,ds$ for $t \in [t_i, q(t_i)]$, $i = 1, 2$.

H3.5.11 $\quad z^2(t) - \varphi^2(t) - z'(t) \ge (2z(t)\varphi(t) - \varphi'(t))\cotg \displaystyle\int_{t_i}^{t} \varphi(s)\,ds$

for $\quad t \in [t_i, q(t_i)]$, $i = 1, 2$.

H3.5.12 $\quad 0 < \displaystyle\int_{t}^{q(t)} \varphi(s)\,ds < \pi$ for $t \in [t_1, t_2]$.

Moreover, assume that the function $r(t)$ satisfies the following condition:
H3.5.13 $\quad r(t) \le t,\ r'(t) > 0,\ r''(t) \le 0$ for $t \in [t_1, q(t_2)]$.
Introduce the following notation:

$$\tilde{a}(t) = \varphi^2(t) + z'(t) + c'(t) - z^2(t) - c(t)z(t)$$

$$+ \cotg \int_{t}^{q(t)} \varphi(s)\,ds(-c(t)\varphi(t) - 2z(t)\varphi(t) + \varphi(t))$$

$$+ \frac{r''(t)}{r'(t)}\left(2z(t) + c(t) + 2\varphi(t)\cotg \int_{t}^{q(t)} \varphi(s)\,ds\right)$$

$$+ [q'^2(t)(2\lambda'(q(t))\varphi(q(t)) + \lambda(q(t))(\varphi'(q(t)) - 2z(q(t))$$

$$\times \varphi(q(t)))) + q''(t)\varphi(q(t))\lambda(q(t)) + q'(t)\mu(q(t))\varphi(q(t))]$$

$$\times \frac{\exp\left(-\int_{t}^{q(t)} z(s)\,ds\right)}{r'(t)\sin \int_{t}^{q(t)} \varphi(s)\,ds}, \qquad t \in [t_1, t_2], \tag{3.5.13}$$

$$\tilde{b}(q(t)) = \frac{\exp\left(\int_t^{q(t)} z(s)\,ds\right)}{\sin\int_t^{q(t)} \varphi(s)\,ds}$$

$$\times \left[r'(t)(c(t)\varphi(t) + 2z(t)\varphi(t) - \varphi'(t)) - 2r''(t)\varphi(t)\right]$$

$$+ q'^2(t)\left[\lambda(q(t))(\varphi^2(q(t)) - z^2(q(t)) + z'(q(t)))\right.$$

$$+ \cotg \int_t^{q(t)} \varphi(s)\,ds(2z(q(t))\varphi(q(t)) - \varphi'(q(t)))$$

$$+ 2\lambda'(q(t))\left(z(q(t)) - \varphi(q(t))\cotg \int_t^{q(t)} \varphi(s)\,ds\right)$$

$$\left. - \lambda''(q(t))\right]$$

$$+ q'(t)\left[\mu(q(t))\left(z(q(t)) - \varphi(q(t))\cotg \int_t^{q(t)} \varphi(s)\,ds\right)\right.$$

$$\left. - \mu'(q(t))\right]$$

$$+ q''(t)\left[\lambda(q(t))\left(z(q(t)) - \varphi(q(t))\cotg \int_t^{q(t)} \varphi(s)\,ds - \lambda'q(t))\right)\right],$$

$$t\in[t_1, t_2]. \tag{3.5.14}$$

Lemma 3.5.1

Assume conditions H3.5.1–H3.5.13 fulfilled, where the functions $\tilde{a}(t)$ and $\tilde{b}(q(t))$ are defined by (3.5.13) and (3.5.14). Then the differential inequality (3.5.2) has a solution $y(t)$ which satisfies conditions (3.5.4)–(3.5.6) as well as the condition

$$(-1)^i y'(t) \leq 0, \quad (-1)^i y''(t) \leq 0 \quad \text{for} \quad t\in[t_i, q(t_i)], \quad i = 1, 2. \tag{3.5.15}$$

Proof

It is immediately verified that the function

$$y(t) = \sin \int_{t_1}^{t} \varphi(s)\, ds \, \exp\left(-\int_{t_1}^{t} z(s)\, ds \right), \quad t \in [t_1, q(t_2)]$$

satisfies inequality (3.5.2). We shall note that conditions H3.5.7–H3.5.9 imply conditions (3.5.4)–(3.5.6) of Theorem 3.5.1 and conditions H3.5.10–H3.5.11 imply condition (3.5.15) of Lemma 3.5.1.

A corollary of Theorem 3.5.1 and Lemma 3.5.1 is the following Theorem 3.5.2 which provides a condition of existence of at least one zero of the solutions of equation (3.5.12) in a given interval.

Theorem 3.5.2

Assume conditions H3.5.1–H3.5.13 fulfilled as well as conditions (2), (4) and (5) of Theorem 3.5.1, where the functions $\tilde{a}(t)$ and $\tilde{b}(q(t))$ are defined by (3.5.13) and (3.5.14), and the following condition

$$\lambda^{(k)}(t) \geq 0, \quad \mu^{(m)}(t) \geq 0 \quad \text{for} \quad t \in (t_1, q(t_1)) \cup (t_2, q(t_2)),$$
$$(3.5.16)$$
$$k = 0, 1, 2; \quad m = 0, 1.$$

Then each solution $x(t)$ of equation (3.5.12) has at least one zero in the interval $[r(t_1), t_2]$.

Consider a particular case of equation (3.5.12) for which the conditions of Theorem 3.5.2 are more comprehensible.

Corollary 3.5.3 *Assume the following conditions fulfilled:*
(1) $r(t) = t - 1$; $\varphi(t) \equiv v \in (0, \pi)$; $c(t) \equiv c = \text{const.} > 0$; $\lambda(t) \equiv \lambda = \text{const.} > 0$; $\mu(t) \equiv \mu = \text{const.} > 0$; $z(t) \equiv k = \text{const.}$
(2) $\mu < c$; $v \, \text{tg} \, v/2 < c/2$.
(3) $k \in [k_v, -v \, \text{tg} \, v/2]$, where k_v is a solution of the inequality

$$e^k \frac{v}{\sin v} (2k_v + c) + \lambda(v^2 - k_v^2 + 2k_v v \cot g \, v) + \mu(k_v - v \cot g \, v) \geq 0$$
$$(3.5.17)$$

$$(4) \quad a(t) \geq c_1(v) \equiv -\lambda e^{-k} 2k \frac{v}{\sin v} + \mu e^{-k} \frac{v}{\sin v} + v^2 - k^2 - ck$$

$$- v \cotg v(2k + c), \quad t \in [t_1, t_1 + \pi v^{-1}] \tag{3.5.18}$$

$$(5) \quad b(t) \geq \begin{cases} 0, \quad t \in [t_1, t_1 + 1] \\ c_2(v) \equiv e^k \dfrac{v}{\sin v}(2k + c) + \mu(k - v \cotg v) \\ \qquad + \lambda^2(v^2 - k^2 + 2kv \cotg v), \quad t \in [t_1 + 1, t_1 + \pi v^{-1}]. \end{cases}$$

$$\tag{3.5.19}$$

Then each solution $x(t)$ of the equation

$$x''(t) + \lambda x''(t - 1) + cx'(t) - \mu x'(t - 1) + a(t)x(t) + b(t)x(t - 1) = 0 \tag{3.5.20}$$

has at least one zero in the interval $[t_1 - 1, t_1 + \pi v^{-1} + 1]$.

Corollary 3.5.4 *Assume the following conditions fulfilled:*
(1) Conditions (1) and (2) of Corollary 3.5.3 hold.
(2) k_0 is a root of the equation

$$e^{k_0}(2k_0 + c) + \mu(k_0 - 1) + \lambda k_0(2 - k_0) = 0 \tag{3.5.21}$$

$$(3) \quad a(t) \geq a_0 > c_1(0) \equiv e^{-k}(\mu - 2k\lambda) - (k^2 + k(c + 2) + c) \tag{3.5.22}$$

for $t \in [t_0, \infty)$, $k \in [k_0, 0]$.

$$(4) \quad b(t) \geq b_0 > c_2(0) \equiv e^k(2k + c) + \mu(k - 1) + \lambda k(2 - k) \tag{3.5.23}$$

for $t \in [t_0, \infty)$, $k \in [k_0, 0]$.

Then each solution $x(t)$ of equation (3.5.20) oscillates.

Remark 3.5.1 Assumptions (3.5.22) and (3.5.23) in Corollary 3.5.4 cannot be improved in the following sense. If in equation (3.5.20) we set $a(t) \equiv c_1$ and $b(t) \equiv c_2(0)$, then it is immediately verified that the function $x(t) = e^{kt}$ is a nonoscillating solution of (3.5.20).

3.6   Notes and comments to Chapter 3

The results of Section 3.1 are those of Grammatikopoulos, Ladas and Meinmaridou [50]. We shall also note the work [49] by the same authors.

The results of Section 3.2 are due to Ladas and Partheniadis [107]. A generalization of these results was obtained by Ladas *et al* [108]. Sufficient conditions for oscillation and asymptotic behaviour of the solutions of second order neutral differential equations with constant coefficients were obtained by Grammatikopoulos, Grove and Ladas [46].

The results of Section 3.3 are due to Bainov *et al* [12]. They generalize some results obtained in the case of concentrated delay.

The results of Section 3.4 are due to Graef *et al* [43]. We shall note that the first oscillation criterion for second order equations, valid both for linear and nonlinear equations, was suggested by Zahariev and Bainov [160]. Sufficient conditions for oscillation of the solutions of second order nonlinear equations were also obtained by Zahariev and Bainov [158] and Bainov *et al* [9]. Oscillatory properties of the solutions of a class of neutral differential equations with 'maxima' were investigated by Bainov and Zahariev [14]. Sufficient conditions for oscillation and asymptotic behaviour of the solutions of second order nonlinear neutral differential equations were obtained by Erbe and Zhang [31] and Grace and Lalli [40], [41].

The results of Section 3.5 are due to Domshlak and Sheikhzamanova [24]. We shall note that the first result on the distribution of the zeros of the solutions for the first order neutral differential equations was obtained by Domshlak [22]. The case of distributed delay for second order neutral differential equations was considered by Sheikhzamanova [140].

4

nth order neutral ordinary differential equations

4.1 *n*th order linear differential equations

In this section the oscillatory properties and asymptotic behaviour of the solution of neutral differential equations of the form

$$\frac{\mathrm{d}^n}{\mathrm{d}t^n}[y(t) + P(t)y(t - \tau)] + Q(t)y(t - \sigma) = 0, \quad t \geq t_0, \quad (4.1.1)$$

are investigated, where $P(t), Q(t) \in C([t_0, \infty); \mathbb{R})$ and the delays τ and σ are nonnegative real numbers.

Let $\varphi(t) \in C([t_0 - \rho, t_0]; \mathbb{R})$, where $\rho = \max\{\tau, \sigma\}$ is a given function, and let $z_k, k = 0, 1, \ldots, n - 1$ be given constants.

Definition 4.1.1 The function $y(t) \in C([t_0 - \rho, \infty); \mathbb{R})$ is said to be a solution of equation (4.1.1) if

$$y(t) = \varphi(t) \quad \text{for} \quad t \in [t_0 - \rho, t_0],$$

$$\frac{\mathrm{d}^k}{\mathrm{d}t^k}[y(t) + P(t)y(t - \tau)]|_{t=t_0} = z_k, \quad k = 0, 1, \ldots, n - 1$$

the function $y(t) + P(t)y(t - \tau)$ is n times continuously differentiable for $t \geq t_0$ and $y(t)$ satisfies equation (4.1.1) for $t \geq t_0$.

We shall note that theorems of existence and uniqueness of the solution of neutral differential equations were obtained by Driver [26], [27], Bellman and Cooke [16] and Hale [66].

We shall first consider the asymptotic behaviour of the nonoscillating solutions of equation (4.1.1).

We shall say that conditions (H4.1) are met if the following conditions hold:

H4.1.1 $P(t) \in C([t_0, \infty); \mathbb{R})$

$$p_1 \le P(t) \le p_2 \quad \text{for} \quad t \in [t_0, \infty)$$

where p_1 and p_2 are constants.

H4.1.2 $Q(t) \in C([t_0, \infty); \mathbb{R})$,

$$Q(t) \ge q > 0 \quad \text{for} \quad t \in [t_0, \infty).$$

Let $y(t)$ be a solution of equation (4.1.1). Set

$$z(t) = y(t) + P(t)y(t - \tau).$$

The following lemma describes some asymptotic properties of the function $z(t)$ when $y(t)$ is a nonoscillating solution of equation (4.1.1).

Lemma 4.1.1

Assume conditions H4.1.1 and H4.1.2 fulfilled. Let $y(t)$ be an eventually positive solution of equation (4.1.1). Set

$$z(t) = y(t) + P(t)y(t - \tau). \tag{4.1.2}$$

Then the following statements are valid:

(i) For each $i = 0, 1, \ldots, n - 1$ the function $z^{(i)}(t)$ is strictly monotone and either

$$\lim_{t \to \infty} z^{(i)}(t) = -\infty \tag{4.1.3}$$

or

$$\lim_{t \to \infty} z^{(i)}(t) = 0 \quad \text{and} \quad z^{(i)}(t)z^{(i+1)}(t) < 0. \tag{4.1.4}$$

(ii) For n even the function $z(t)$ is negative.

(iii) Assume that $p_2 < -1$ and that n is odd. Then (4.1.3) holds.

(iv) Assume that $p_1 \ge -1$ and that n is odd or even. Then (4.1.4) holds, and in particular, $z(t)$ is bounded.

Proof

(*i*) From equation (4.1.1) we find

$$z^{(n)}(t) = -Q(t)y(t - \sigma) \leqq -qy(t - \sigma) < 0 \qquad (4.1.5)$$

which implies that $z^{(n-1)}(t)$ is a strictly decreasing function of t, while $z^{(i)}(t)$, $i = 0, 1, \ldots, n - 2$ are strictly monotone functions of t. Therefore either

$$\lim_{t \to \infty} z^{(n-1)}(t) = -\infty \qquad (4.1.6)$$

or

$$\lim_{t \to \infty} z^{(n-1)}(t) \equiv l \in \mathbb{R}. \qquad (4.1.7)$$

Assume that (4.1.6) is satisfied. Then it is easily seen that (4.1.3) holds. Assume now that (4.1.7) holds. Then integrating both sides of (4.1.5) from t_1 to t, with t_1 large enough, and letting $t \to \infty$ we find

$$\int_{t_1}^{\infty} qy(t - \sigma) \, ds \leqq z^{(n-1)}(t_1) - l$$

which implies that $y \in L_1[t_1, \infty)$. Thus, in view of H4.1.1, $z \in L_1[t_1, \infty)$. Since $z(t)$ is monotone, it follows that

$$\lim_{t \to \infty} z(t) = 0 \qquad (4.1.8)$$

and so also $l = 0$. Finally, from (4.1.8) we conclude that consecutive derivatives of $z(t)$ alternate sign, that is, for each $i = 0, 1, \ldots, n - 1$

$$z^{(i)}(t)z^{(i+1)}(t) < 0.$$

(*ii*) Clearly for n even, either (4.1.3) or (4.1.4) implies that $z(t) < 0$.

(*iii*) If (4.1.3) were false, then (4.1.4) would hold and so

$$z(t) > 0. \qquad (4.1.9)$$

Therefore

$$y(t) > -P(t)y(t - \tau) \geqq -p_2 y(t - \tau)$$

and by iteration

$$y(t + k\tau) > (-p_2)^k y(t) \to \infty \quad \text{as} \quad k \to \infty.$$

Hence

$$\lim_{t \to \infty} y(t) = \infty$$

and (4.1.5) implies that

$$\lim_{t \to \infty} z^{(n)}(t) = -\infty.$$

Thus

$$\lim_{t \to \infty} z(t) = -\infty$$

which contradicts (4.1.9) and proves (4.1.3).
    (*iv*) If (4.1.4) were false, then, from (4.1.3),

$$\lim_{t \to \infty} z(t) = -\infty \qquad (4.1.10)$$

and so

$$z(t) < 0.$$

Hence

$$y(t) < -P(t)y(t - \tau) \leqq -p_1 y(t - \tau) \leqq y(t - \tau)$$

which implies that $y(t)$ is a bounded function. This contradicts (4.1.10). The proof of the lemma is complete.

Using the asymptotic properties of the function $z(t)$, we obtain the following result about the asymptotic behaviour of the nonoscillating solutions of equation (4.1.1).

Theorem 4.1.1

Consider the neutral delay differential equation (4.1.1) and assume that conditions H4.1.1 and H4.1.2 are satisfied. Then the following statements are true:

(a) *Assume that*

$$n \text{ is odd} \quad \text{and} \quad p_2 < -1.$$

Then every nonoscillating solution $y(t)$ of equation (4.1.1) tends to $+\infty$ or $-\infty$ as $t \to \infty$.
   (b) *Assume that*

$$n \text{ is odd} \quad \text{and} \quad p_1 \geq 0 \qquad\qquad (4.1.11)$$

or

$$(n \text{ is even or odd} \quad \text{and}) \quad -1 < p_1 \leq p_2 < 0. \qquad (4.1.12)$$

Then every nonoscillating solution $y(t)$ of equation (4.1.1) tends to zero as $t \to \infty$.

Proof

As the negative of a solution of equation (4.1.1) is also a solution of the same equation, it suffices to prove the theorem for an eventually positive solution $y(t)$ of equation (4.1.1).
   (a) Set

$$z(t) = y(t) + P(t)y(t - \tau).$$

Then from Lemma 4.1.1 (*iii*) we have

$$\lim_{t \to \infty} z(t) = -\infty.$$

Observe that

$$p_1 y(t - \tau) \leq P(t)y(t - \tau) < z(t) \to -\infty \quad \text{as} \quad t \to \infty$$

and so

$$\lim_{t \to \infty} y(t) = \infty.$$

(*b*) Assume that (4.1.11) holds. Then, using Lemma 4.1.1 (*iv*), we find

$$0 < y(t) \leq z(t) \to 0 \quad \text{as} \quad t \to \infty$$

and so

$$\lim_{t \to \infty} y(t) = 0.$$

Next assume that (4.1.12) is satisfied. Then Lemma 4.1.1 (*iv*) implies that (4.1.4) holds. Depending on whether n is even or odd, we distinguish the following two cases:

Case 1. n is even. In this case Lemma 4.1.1 (*ii*) implies that

$$z(t) < 0$$

and hence

$$y(t) < -P(t)y(t - \tau) < y(t - \tau).$$

Therefore, $y(t)$ is a bounded function. Assume, for the sake of contradiction, that

$$\limsup_{t \to \infty} y(t) \equiv s > 0.$$

Let $\{t_k\}$ be a sequence of points such that

$$\lim_{k \to \infty} t_k = \infty \quad \text{and} \quad \lim_{k \to \infty} y(t_k) = s.$$

Then, for k large enough,

$$z(t_k) = y(t_k) + P(t_k)y(t_k - \tau) \geq y(t_k) + p_1 y(t_k - \tau)$$

and hence

$$\limsup_{t \to \infty} y(t_k - \tau) \geq s/(-p_1) > s$$

which is a contradiction.

Case 2. n is odd. First we shall prove that $y(t)$ is a bounded function. To this end, observe that from Lemma 4.1.1 (*iv*) we have

$$z(t) > 0, \quad z'(t) < 0 \quad \text{and} \quad \lim_{t \to \infty} z(t) = 0.$$

Therefore, there is a positive constant B such that

$$z(t) \leq B$$

and so

$$y(t) < -P(t)y(t - \tau) + B \leq -p_1 y(t - \tau) + B. \qquad (4.1.13)$$

Assume, for the sake of contradiction, that $y(t)$ is not bounded. Then there exists a sequence of points $\{t_k\}$ such that

$$\lim_{k \to \infty} t_k = \infty, \quad \lim_{k \to \infty} y(t_k) = \infty \quad \text{and} \quad y(t_k) = \max_{t_0 \leq s \leq t_k} y(s).$$

Thus, from (4.1.13) we have

$$y(t_k) < -p_1 y(t_k - \tau) + B \leq -p_1 y(y_k) + B$$

or equivalently

$$(1 + p_1)y(t_k) \leq B$$

which as $k \to \infty$ leads to a contradiction. Set

$$s \equiv \lim_{t \to \infty} \sup y(t)$$

which exists because $y(t)$ is bounded. Let $\{t_k\}$ be a sequence of points such that

$$\lim_{k \to \infty} t_k = \infty \quad \text{and} \quad \lim_{k \to \infty} y(t_k) = s.$$

Then, for k large enough,

$$z(t_k) = y(t_k) + P(t_k)y(t_k - \tau) \geq y(t_k) + p_1 y(t_k - \tau)$$

which as $k \to \infty$ implies that $s = 0$. The proof of the theorem is complete.

The following examples show that Theorem 4.1.1 may be not true if either condition H4.1.1 or H4.1.2 is not satisfied.

Example 4.1.1   In the neutral delay differential equation

$$\frac{d^n}{dt^n}[y(t) - e^t y(t - 1)] + e^{-2} y(t - 2) = 0, \quad t \geq 1, \quad n \text{ is odd}$$

condition H4.1.1 of Theorem 4.1.1 (a) is not satisfied. Note that $y(t) = e^{-t}$ is a solution of this equation with $\lim_{t \to \infty} y(t) = 0$.

Example 4.1.2 In the neutral delay differential equation

$$\frac{d^2}{dt^2}[y(t) + (-\tfrac{1}{2} + (t-1)^{-\frac{1}{2}})y(t-1)]$$

$$+ \tfrac{1}{4}(t-2)^{-\frac{1}{2}}(t^{-\frac{3}{2}} - \tfrac{1}{2}(t-1)^{-\frac{3}{2}})y(t-2) = 0, \quad t \geq 3$$

condition H4.1.2 of Theorem 4.1.1 (b) is not satisfied. Note that $y(t) = \sqrt{t}$ is a solution of this equation with $\lim_{t \to \infty} y(t) = \infty$.

In the subsequent theorems sufficient conditions are obtained for oscillation of the solutions of equation (4.1.1) in the case when n is odd.

The following lemma which will be used in the proofs of Theorems 4.1.2, 4.1.3 and 4.1.6 has been extracted from results due to Ladas and Stavroulakis [104], [106].

Lemma 4.1.2

Assume that n is odd and r and μ are positive constants such that

$$r^{1/n}\frac{\mu}{n} > \frac{1}{e}.$$

Then, the following statements are true:
   (i) the inequality

$$x^{(n)}(t) - rx(t + \mu) \leq 0$$

has no eventually negative solution;
   (ii) the inequality

$$x^{(n)}(t) - rx(t + \mu) \geq 0$$

has no eventually positive solution;
   (iii) the inequality

$$x^{(n)}(t) + rx(t - \mu) \geq 0$$

has no eventually negative solution;

(iv) the inequality

$$x^{(n)}(t) + rx(t - \mu) \le 0$$

has no eventually positive solution.

Let $y(t)$ be a solution of equation (4.1.1). Set

$$z(t) = y(t) + P(t)y(t - \tau).$$

Then a direct substitution shows that $z(t)$ is an n times continuously differentiable solution of the neutral delay differential equation

$$z^{(n)}(t) + R(t)z^{(n)}(t - \tau) + Q(t)z(t - \sigma) = 0, \quad t \ge t_0 \qquad (4.1.14)$$

where

$$R(t) = P(t - \sigma)\frac{Q(t)}{Q(t - \tau)}.$$

Theorems 4.1.2, 4.1.3 and 4.1.6 below provide sufficient conditions for the oscillation of all solutions, while Theorems 4.1.4 and 4.1.5 deal with unbounded solutions only.

Theorem 4.1.2

Consider the neutral delay differential equation (4.1.1) and assume that n is odd and that conditions H4.1.1 and H4.1.2 are satisfied with

$$p_2 < -1.$$

Suppose also that there exists a positive constant r such that $\tau > \sigma$,

$$\frac{Q(t)}{P(t + \tau - \sigma)} \le -r \qquad (4.1.15)$$

and

$$r^{1/n}\frac{\tau - \sigma}{n} > \frac{1}{e}. \qquad (4.1.16)$$

Then each solution of equation (4.1.1) oscillates.

Proof

Otherwise there exists an eventually positive solution $y(t)$ of equation (4.1.1). Set

$$z(t) = y(t) + P(t)y(t - \tau).$$

Then

$$z^{(n)}(t) = -Q(t)y(t - \sigma) \leq 0$$

and also, Lemma 4.1.1 *(iii)* implies that

$$z(t) < 0.$$

As $z^{(n)}(t) \leq 0$, from equation (4.1.14) it follows that

$$z^{(n)}(t) + \frac{Q(t)}{P(t + \tau - \sigma)} z(t + \tau - \sigma) \leq 0$$

which, by Lemma 4.1.2 *(i)*, (4.1.15) and (4.1.16), has no eventually negative solution. This is a contradiction. The proof is complete.

Theorem 4.1.3

Consider the neutral delay differential equation (4.1.1) and assume that n is odd and that the conditions H4.1.1 and H4.1.2 are satisfied with

$$-1 \leq p_1 \leq p_2 \leq 0.$$

Suppose also that

$$q^{1/n} \frac{\sigma}{n} > \frac{1}{e}. \tag{4.1.17}$$

Then every solution of equation (4.1.1) oscillates.

Proof

Otherwise there exists an eventually positive solution $y(t)$ of equation (4.1.1). Set

$$z(t) = y(t) + P(t)y(t - \tau).$$

Then $z(t)$ is a solution of equation (4.1.14). Also

$$z^{(n)}(t) < 0$$

and from Lemma 4.1.1 (*iv*)

$$z(t) > 0 \qquad\qquad (4.1.18)$$

Then, from equation (4.1.14) we find

$$z^{(n)}(t) + qz(t - \sigma) < 0.$$

But, because of Lemma 4.1.2 (*iv*) and (4.1.17), it is impossible for this inequality to have an eventually positive solution. This contradicts (4.1.18). The proof of the theorem is complete.

In Theorems 4.1.4 and 4.1.5 below condition H4.1.2 is not required.

Theorem 4.1.4

Consider the neutral delay differential equation (4.1.1) and assume that n is odd,

$$-1 < p_1 \leq P(t) \leq 0,$$

$$Q(t) \geq 0$$

and

$$\int_{t_0}^{\infty} Q(s)\,ds = \infty. \qquad\qquad (4.1.19)$$

Then each unbounded solution of equation (4.1.1) oscillates.

Proof

Otherwise equation (4.1.1) has an eventually positive unbounded solution $y(t)$. Set

$$z(t) = y(t) + P(t)y(t - \tau).$$

Then

$$z^{(n)}(t) = -Q(t)y(t - \sigma) \leq 0. \qquad\qquad (4.1.20)$$

We shall prove that

$$z(t) > 0 \tag{4.1.21}$$

and

$$z'(t) > 0. \tag{4.1.22}$$

If (4.1.21) were false, then $z(t) \leqq 0$, which implies that

$$y(t) \leqq -P(t)y(t - \tau) \leqq y(t - \tau)$$

that is, $y(t)$ is unbounded. On the other hand, if (4.1.22) were false, then $z'(t) \leqq 0$ and, in particular, $z(t)$ would be a bounded function. Thus, there would exist a positive constant B such that

$$y(t) \leqq -P(t)y(t - \tau) + B \leqq -p_1 y(t - \tau) + B. \tag{4.1.23}$$

As $y(t)$ is unbounded, there must exist a sequence of points $\{t_k\}$ such that

$$\lim_{k \to \infty} t_k = \infty, \quad \lim_{k \to \infty} y(t_k) = \infty \quad \text{and} \quad y(t_k) = \max_{t_0 \leqq s \leqq t_k} y(s).$$

Then, from (4.1.23)

$$y(t_k) \leqq -p_1 y(t_k) + B$$

or equivalently

$$(1 + p_1)y(t_k) \leqq B$$

which contradicts the fact that $y(t_k)$ is unbounded. Thus (4.1.21) and (4.1.22) have been established. Since

$$0 < z(t) < y(t)$$

equation (4.1.20) implies that

$$z^{(n)}(t) + Q(t)z(t - \sigma) \leqq 0.$$

Integrating from t_1 to t, for t_1 large enough, we find

$$z^{(n-1)}(t) - z^{(n-1)}(t_1) + z(t_1 - \sigma) \int_{t_1}^{t} Q(s) \, ds \leqq 0$$

which, in view of (4.1.19), implies that

$$z^{(n-1)}(t) < 0. \qquad (4.1.24)$$

From (4.1.24) and (4.1.20) it follows that $z(t) < 0$, which contradicts (4.1.21). The proof is complete.

Theorem 4.1.5

Consider the neutral delay differential equation (4.1.1) and assume that n is odd,

$$0 \leqq P(t) \leqq 1, \quad Q(t) \geqq 0$$

and

$$\int_{t_0}^{\infty} Q(s)[1 - P(s - \sigma)] \, ds = \infty.$$

Then each unbounded solution of equation (4.1.1) oscillates.

Proof

Otherwise, there exists an eventually positive unbounded solution $y(t)$ of equation (4.1.1). Set

$$z(t) = y(t) + P(t)y(t - \tau).$$

Then

$$z^{(n)}(t) = -Q(t)y(t - \sigma) \leqq 0. \qquad (4.1.25)$$

Clearly

$$z(t) > 0, \quad z(t) > y(t) \quad \text{and} \quad z(t) \text{ is unbounded.}$$

This is already impossible for $n = 1$, while for $n \geqq 3$ we have $z'(t) > 0$ and so

$$z(t - \sigma) < y(t - \sigma) + P(t - \sigma)z(t - \tau - \sigma) < y(t - \sigma) + P(t - \sigma)z(t - \sigma)$$

or

$$y(t - \sigma) > [1 - P(t - \sigma)]z(t - \sigma).$$

Thus, equation (4.1.25) yields

$$z^{(n)}(t) + Q(t)[1 - P(t - \sigma)]z(t - \sigma) \leqq 0.$$

Integrating from t_1 to t, for t_1 large enough, we find

$$z^{(n-1)}(t) - z^{(n-1)}(t_1) + z(t_1 - \sigma) \int_{t_1}^{t} Q(s)[1 - P(t - \sigma)] \, ds \leqq 0$$

which as $t \to \infty$ leads to a contradiction. The proof is complete.

Remark 4.1.1 For $n = 1$, the conclusion of Theorem 4.1.5 remains true under the hypotheses

$$P(t) \geqq p_1 > -1 \quad \text{and} \quad Q(t) \geqq 0$$

only. The proof of this follows from Theorem 4.1.3 in Chapter 2.

Theorem 4.1.6

Consider the neutral delay differential equation (4.1.1) and assume that n is odd, H4.1.2 holds, $Q(t)$ is a τ-periodic function and

$$P(t) \equiv p \in \mathbb{R}.$$

Then each of the following conditions implies that every solution of equation (4.1.1) oscillates:

(i) $p < -1, \tau > \sigma$ and

$$\left(-\frac{q}{1+p} \right)^{1/n} \frac{\tau - \sigma}{n} > \frac{1}{e}; \tag{4.1.26}$$

(ii) $p = -1;$ \hfill (4.1.27)

(iii) $p > -1, \sigma > \tau$ and

$$\left(\frac{q}{1+p} \right)^{1/n} \frac{\sigma - \tau}{n} > \frac{1}{e}. \tag{4.1.28}$$

Proof

Assume that one of conditions (i)–(iii) is satisfied and that, contrary to the conclusion of the theorem, equation (4.1.1) has an eventually positive solution $y(t)$. Set

$$z(t) = y(t) + py(t - \tau) \quad \text{and} \quad w(t) = z(t) + pz(t - \tau).$$

Then

$$z^{(n)}(t) = -Q(t)y(t - \sigma) \leqq 0,$$

$$w^{(n)}(t) = -Q(t)z(t - \sigma)$$

and

$$w^{(n)}(t) + pw^{(n)}(t - \tau) + Q(t)w(t - \sigma) = 0. \tag{4.1.29}$$

First, assume that (4.1.26) is satisfied. Then by Lemma 4.1.1 (iii) it follows that (4.1.3) is satisfied. This implies that

$$w^{(n)}(t) = -Q(t)z(t - \sigma) \geqq -qz(t - \sigma) \to +\infty$$

and so

$$w(t) > 0. \tag{4.1.30}$$

Also

$$w^{(n)}(t - \tau) = -Q(t)z(t - \tau - \sigma) \leqq -Q(t)z(t - \sigma) = w^{(n)}(t). \tag{4.1.31}$$

Substituting (4.1.31) into equation (4.1.29), we find

$$w^{(n)}(t) + \frac{q}{1 + p} w(t + \tau - \sigma) \geqq 0. \tag{4.1.32}$$

But, in view of Lemma 4.1.2 (ii) and (4.1.26), inequality (4.1.32) cannot have an eventually positive solution. This contradicts (4.1.30) and proves the theorem when (4.1.26) is satisfied.

Next, assume that (4.1.27) holds. Then, by Lemma 4.1.1 (iv) it follows that (4.1.4) is satisfied. This implies that (4.1.31) is true and that

$$w(t) > 0.$$

Hence (4.1.29) yields

$$Q(t)w(t - \sigma) \leqq 0$$

which contradicts the fact that $w(t)$ is positive.

Finally, assume that (4.1.28) is satisfied. Then, again, (4.1.31) holds and (4.1.29) implies that

$$w^{(n)}(t) + \frac{q}{1 + p} w(t - (\sigma - \tau)) \leqq 0. \qquad (4.1.33)$$

But, in view of Lemma 4.1.2 (iv) and (4.1.28), inequality (4.1.33) cannot have an eventually positive solution. This contradicts (4.1.30). The proof of the theorem is complete.

In the subsequent theorems sufficient conditions are obtained for oscillation of the solutions of equation (4.1.1) in the case when n is even.

The following lemma, which will be used in the proofs of Theorem 4.1.8 and 4.1.11 below, has been extracted from results due to Ladas and Stavroulakis [104].

Lemma 4.1.3

Assume that n is even and r and μ are positive constants such that

$$r^{1/n} \frac{\mu}{n} > \frac{1}{e}.$$

Then the inequality

$$x^{(n)}(t) - rx(t - \mu) \leqq 0$$

has no eventually negative bounded solution.

Theorems 4.1.7–4.1.10 below provide sufficient conditions for the oscillation of all solutions of equation (4.1.1), while Theorems 4.1.11 and 4.1.12 deal with the oscillation of all bounded and all unbounded solutions respectively.

Theorem 4.1.7

Consider the neutral delay differential equation (4.1.1) and assume that n is even and that conditions H4.1.1 and H4.1.2 hold. Furthermore assume that P(t) is not eventually negative. Then every solution of equation (4.1.1) oscillates.

Proof

Assume, for the sake of contradiction, that $y(t)$ is an eventually positive solution of equation (4.1.1). Set

$$z(t) = y(t) + P(t)y(t - \tau).$$

Then, eventually, $z(t)$ takes nonnegative values. But, since n is even, Lemma 4.1.1 (*ii*) implies that $z(t)$ is eventually negative. This contradiction completes the proof.

The example below illustrates Theorem 4.1.7.

Example 4.1.3 The neutral delay differential equation

$$\frac{d^2}{dt^2}[y(t) + (\tfrac{1}{2} + \cos t)y(t - 2\pi)] + (\tfrac{3}{2} + \cos t)y(t - 4\pi) = 0, \quad t \geq 0$$

satisfies the hypotheses of Theorem 4.1.7. Therefore, every solution of this equation oscillates. For instance,

$$y(t) = \frac{\cos t}{3/2 + \cos t}$$

is an oscillating solution.

The following example shows that if we remove condition H4.1.2 from Theorem 4.1.7, the result may be not true.

Example 4.1.4 The neutral delay differential equation

$$\frac{d^2}{dt^2}[y(t) + (t - 1)^{-1/2}y(t - 1)] + \tfrac{1}{4}t^{-3/2}(t-2)^{-1/2}y(t - 2) = 0, \quad t > 2,$$

satisfies all the hypotheses of Theorem 4.1.7 except for H4.1.2. Note that $y(t) = \sqrt{t}$ is a nonoscillating solution of this equation.

Theorem 4.1.8

Consider the neutral delay differential equation (4.1.1) and assume that n is even and that conditions H4.1.1 and H4.1.2 hold with

$$-1 \leq p_1 \leq p_2 < 0.$$

Suppose also that there exists a positive constant r such that $\sigma > \tau$,

$$\frac{Q(t)}{P(t + \tau - \sigma)} \leqq -r \tag{4.1.34}$$

and

$$r^{1/n} \frac{\sigma - \tau}{n} > \frac{1}{e}. \tag{4.1.35}$$

Then every solution of equation $(4.1.1)$ oscillates.

Proof

Otherwise there exists an eventually positive solution $y(t)$ of equation $(4.1.1)$. Set

$$z(t) = y(t) + P(t)y(t - \tau).$$

Then $z(t)$ is a solution of equation $(4.1.14)$. Clearly

$$z^{(n)}(t) < 0 \tag{4.1.36}$$

and from Lemma 4.1.1 (ii) and (iv) we conclude that $z(t)$ is an eventually negative and bounded function. Using $(4.1.36)$, from equation $(4.1.14)$ we obtain

$$R(t)z^{(n)}(t - \tau) + Q(t)z(t - \sigma) > 0.$$

Hence

$$z^{(n)}(t) + \frac{Q(t)}{P(t + \tau - \sigma)} z(t - (\sigma - \tau)) < 0$$

which, in view of $(4.1.34)$, leads to the inequality

$$z^{(n)}(t) - rz(t - (\sigma - \tau)) < 0.$$

But, because of $(4.1.35)$, Lemma 4.1.3 implies that it is impossible for this inequality to have an eventually negative bounded solution. This completes the proof of the theorem.

Example 4.1.5   For the neutral delay differential equation

$$\frac{d^2}{dt^2}[y(t) - (4 + e^{-t})y(t-1)] + e(4-e)y(t-2) = 0, \quad t \geq 1.$$

the hypothesis that $-1 \leq p_1$ is not satisfied. Note that $y(t) = e^t$ is a nonoscillating solution of this equation.

In Theorems 4.1.9 and 4.1.10 condition H4.1.2 is not required.

Theorem 4.1.9

Consider the neutral delay differential equation (4.1.1) and assume that n is even. Suppose also that

$$Q(t) \geq 0, \quad Q(t) \not\equiv 0 \quad \text{and} \quad \tau\text{-periodic,}$$

$$0 \leq P(t) \equiv p \text{ is constant.}$$

Then every solution of equation (4.1.1) oscillates.

Proof

Otherwise there exists an eventually positive solution $y(t)$ of equation (4.1.1). Set

$$z(t) = y(t) + py(t - \tau) \quad \text{and} \quad w(t) = z(t) + pz(t - \tau).$$

Then

$$z(t) > 0 \quad \text{and} \quad w(t) > 0.$$

Also

$$z^{(n)}(t) = -Q(t)y(t - \sigma) \leq 0$$

and

$$w^{(n)}(t) = -Q(t)z(t - \sigma) \leq 0.$$

We claim that

$$z^{(n-1)}(t) \geq 0 \quad \text{and} \quad w^{(n-1)}(t) \geq 0. \tag{4.1.37}$$

Otherwise

$$z^{(n-1)}(t) < 0 \quad \text{or} \quad w^{(n-1)}(t) < 0$$

which together with $z^{(n)}(t) \leq 0$ and $w^{(n)}(t) \leq 0$ implies that

$$z(t) < 0 \quad \text{or} \quad w(t) < 0$$

which is a contradiction.

Next, we claim that

$$z'(t) \geq 0 \quad \text{and} \quad w'(t) \geq 0.$$

Otherwise

$$z'(t) < 0 \quad \text{or} \quad w'(t) < 0.$$

But any one of these inequalities implies that the higher derivatives of odd order of that function are also negative. This contradicts (4.1.37). Thus we have proved that $z(t)$ and $w(t)$ are increasing functions of t. Observe now that $w(t)$ is a continuously differentiable solution of the neutral delay differential equation

$$w^{(n)}(t) + pw^{(n)}(t - \tau) + Q(t)w(t - \sigma) = 0. \tag{4.1.38}$$

As

$$w^{(n)}(t - \tau) = -Q(t)z(t - \sigma - \tau) \geq -Q(t)z(t - \sigma) = w^{(n)}(t)$$

equation (4.1.38) implies that

$$w^{(n)}(t) + \frac{1}{1 + p} Q(t)w(t - \sigma) \leq 0.$$

Integrating both sides of this inequality from t_1 to t, with t_1 large enough, we find that

$$w^{(n-1)}(t) - w^{(n-1)}(t_1) + \frac{1}{1 + p} w(t_1 - \sigma) \int_{t_1}^{t} Q(s)\, ds \leq 0$$

which leads to a contradiction as $t \to \infty$. The proof is complete.

Theorem 4.1.10

Consider the neutral delay differential equation (4.1.1) and assume that n is even. Assume also that

$$0 \leq P(t) \leq 1, \quad Q(t) \geq 0$$

and that

$$\int_{t_1}^{\infty} Q(s)[1 - P(t - \sigma)]\, ds = \infty.$$

Then every solution of equation (4.1.1) oscillates.

Proof

Assume, for the sake of contradiction, that $y(t)$ is an eventually positive solution of equation (4.1.1). Set

$$z(t) = y(t) + P(t)y(t - \tau). \qquad (4.1.39)$$

Then

$$z(t) \geq 0 \qquad (4.1.40)$$

and

$$z^{(n)}(t) \leq 0. \qquad (4.1.41)$$

Hence $z^{(n-1)}(t)$ is a decreasing function of t. We claim that

$$z^{(n-1)}(t) \geq 0. \qquad (4.1.42)$$

Otherwise

$$z^{(n-1)}(t) < 0$$

which together with (4.1.41) implies that

$$\lim_{t \to \infty} z^{(k)}(t) = -\infty, \quad k = 0, 1, \ldots, n - 2.$$

But this contradicts (4.1.40).

Next, observe that from equation (4.1.1) we have

$$z^{(n)}(t) + Q(t)y(t - \sigma) = 0. \tag{4.1.43}$$

Using (4.1.39) in (4.1.43), we see that

$$z^{(n)}(t) + Q(t)[z(t - \sigma) - P(t - \sigma)y(t - \tau - \sigma)] = 0. \tag{4.1.44}$$

As

$$z(t) > y(t)$$

(4.1.44) yields

$$z^{(n)}(t) + Q(t)[z(t - \sigma) - P(t - \sigma)z(t - \tau - \sigma)] \leqq 0$$

which, in view of (4.1.42), leads to the inequality

$$z^{(n)}(t) + Q(t)[1 - P(t - \sigma)]z(t - \sigma) \leqq 0. \tag{4.1.45}$$

Integrating both sides of (4.1.45) from t_1 to t, with t_1 large enough, we find that

$$z^{(n-1)}(t) - z^{(n-1)}(t_1) + z(t_1 - \sigma) \int_{t_1}^{t} Q(s)[1 - P(s - \sigma)] \, ds \leqq 0$$

which, as $t \to \infty$, leads to a contradiction. The proof is complete.

Theorem 4.1.11

Consider the neutral delay differential equation (4.1.1) and assume that n is even. Assume also that conditions H4.1.1 and H4.1.2 are satisfied with

$$p_2 < 0$$

and that there exists a positive number r such that $\sigma > \tau$,

$$\frac{Q(t)}{P(t + \tau - \sigma)} \leqq -r \tag{4.1.46}$$

and

$$r^{1/n} \frac{\sigma - \tau}{n} > \frac{1}{e}.$$

Then every bounded solution of equation (4.1.1) oscillates.

Proof

Otherwise there exists an eventually positive bounded solution $y(t)$ of equation (4.1.1). Set

$$z(t) = y(t) + P(t)y(t - \tau).$$

Then

$$z^{(n)}(t) < 0. \tag{4.1.47}$$

Since n is even,

$$z(t) < 0. \tag{4.1.48}$$

Therefore, $z(t)$ is an eventually negative and bounded solution of equation (4.1.1). In view of (4.1.48), (4.1.47) and (4.1.46), we obtain

$$z^{(n)}(t) - rz(t - (\sigma - \tau)) < 0.$$

But, Lemma 4.1.3 implies that the above inequality has no eventually negative bounded solution. This contradicts (4.1.48) and the proof is complete.

Example 4.1.5, which we presented earlier, also illustrates that under the hypotheses of Theorem 4.1.11, equation (4.1.1) may have unbounded nonoscillating solutions.

In the next result condition H4.1.2 is not required.

Theorem 4.1.12

Consider the neutral delay differential equation (4.1.1). Assume that n is even,

$$-1 \le P(t) \le 0,$$

$$Q(t) \ge 0$$

and that

$$\int_{t_0}^{\infty} Q(s)\, ds = \infty.$$

Then every unbounded solution of equation (4.1.1) oscillates.

Proof

Assume, for the sake of contradiction, that $y(t)$ is an unbounded positive solution of equation (4.1.1). Set

$$z(t) = y(t) + P(t)y(t - \tau).$$

We have

$$z^{(n)}(t) = -Q(t)y(t - \sigma) \leq 0$$

and so $z^{(i)}(t)$, for $i = 0, 1, \ldots, n - 1$, are monotone functions. We claim that

$$z^{(n-1)}(t) \geq 0, \quad z'(t) \geq 0 \quad \text{and} \quad z(t) \geq 0.$$

Otherwise $z(t) < 0$ and so

$$y(t) < -P(t)y(t - \tau) \leq y(t - \tau)$$

which is impossible since $y(t)$ is unbounded. Integrating (4.1.1) from t_1 to t, with t_1 large enough, we find

$$z^{(n-1)}(t) - z^{(n-1)}(t_1) + z(t_1 - \sigma) \int_{t_1}^{t} Q(s)\, ds \leq 0$$

which, as $t \to \infty$, is impossible. The proof is complete.

Example 4.1.6 The neutral delay differential equation

$$\frac{d^2}{dt^2}\left[y(t) - \frac{2\,e^{\pi/2} + e^{-\pi/2}}{e^{3\pi/2} + 2\,e^{-3\pi/2}} y(t - 2\pi) \right] + \frac{2(e^{2\pi} - e^{-2\pi})}{e^{3\pi/2} + 2\,e^{-3\pi/2}} y\left(t - \frac{\pi}{2}\right) = 0,$$

$$t \geq 0$$

satisfies the hypotheses of Theorem 4.1.12. Therefore, every unbounded solution of this equation oscillates. For instance, $y(t) = e^t \cos t$ is such a solution. On the other hand, the bounded solutions of this equation do not have to oscillate. For instance, $y(t) = e^{-t}$ is such a solution.

4.2   *n*th order differential equations with constant coefficients

Consider the *n*th order neutral delay differential equation

$$\frac{d^n}{dt^n} [x(t) - px(t - \tau)] + qx(t - \sigma) = 0 \qquad (4.2.1)$$

where *n* is an odd natural number and

$$0 \leq p \leq 1, \quad q > 0 \quad \text{and} \quad \tau, \sigma > 0. \qquad (4.2.2)$$

Furthermore, assume the following condition fulfilled:
    H4.2.1   There exists a nonnegative integer N such that every solution of the delay differential equation

$$y^{(n)}(t) + \sum_{i=0}^{N} qp^i y(t - \sigma - i\tau) = 0 \qquad (4.2.3)$$

oscillates.

Theorem 4.2.1

Consider the neutral delay differential equation (4.2.1) where n is odd and assume that conditions (4.2.2) and H4.2.1 hold. Then every solution of equation (4.2.1) oscillates.

Proof

Assume, for the sake of contradiction, that equation (4.2.1) has an eventually positive solution $x(t)$. Set

$$z(t) = x(t) - px(t - \tau). \qquad (4.2.4)$$

Then

$$z^{(n)}(t) = -qx(t - \sigma) < 0. \tag{4.2.5}$$

Thus $z^{(n-1)}(t)$ decreases and either

$$\lim_{t \to \infty} z^{(n-1)}(t) = -\infty \tag{4.2.6}$$

or

$$\lim_{t \to \infty} z^{(n-1)}(t) \equiv L \in \mathbb{R}. \tag{4.2.7}$$

We shall prove that (4.2.7) holds and that $L = 0$. In fact, if (4.2.6) were true, then

$$\lim_{t \to \infty} z(t) = -\infty \tag{4.2.8}$$

and, in particular,

$$z(t) < 0$$

which implies that

$$x(t) < px(t - \tau) \leq x(t - \tau).$$

Hence $x(t)$ is bounded which contradicts (4.2.8). Thus (4.2.7) holds. We now claim that $L = 0$. To this end, observe that $z(t)$ is an n-times differentiable solution of (4.2.1). That is,

$$z^{(n)}(t) - pz^{(n)}(t - \tau) + qz(t - \sigma) = 0 \tag{4.2.9}$$

If $L \neq 0$, then $\lim_{t \to \infty} z(t) \neq 0$ and so

$$\lim_{t \to \infty} \frac{d}{dt}[z^{(n-1)}(t) - pz^{(n-1)}(t - \tau)] \neq 0$$

which contradicts (4.2.7). Thus $L = 0$. Next, we shall show that

$$\lim_{t \to \infty} z(t) = 0. \tag{4.2.10}$$

In fact, integrating both sides of (4.2.5) from t_1 to t and letting $t \to \infty$, we find that

$$z^{(n-1)}(t_1) = q \int_{t_1}^{\infty} x(s-\sigma)\,ds$$

which shows that $x \in L_1[t_1, \infty)$. Then $z \in L_1[t_1, \infty)$ and as z is monotone, it follows that (4.2.10) holds. Also as $z^{(n-1)}(t)$ decreases to zero, it follows that

$$z^{(n-1)}(t) > 0 \tag{4.2.11}$$

From (4.2.10) and (4.2.11) and the fact that n is odd it follows that

$$z(t) > 0 \tag{4.2.12}$$

We also have

$$z(t) \leq x(t) \tag{4.2.13}$$

From (4.2.4) and (4.2.5) we see that

$$z^{(n)}(t) + qz(t-\sigma) + qpx(t-\sigma-\tau) = 0$$

and by induction

$$z^{(n)}(t) + \sum_{i=0}^{N} qp^i z(t-\sigma-i\tau) + qp^{N+1}x(t-\sigma-(N+1)\tau) = 0.$$

Hence,

$$z^{(n)}(t) + \sum_{i=0}^{N} qp^i z(t-\sigma-i\tau) \leq 0. \tag{4.2.14}$$

From (4.2.7), with $L = 0$, and (4.2.10) it follows that for $i = 1, 2, \ldots, n-1$

$$\lim_{t \to \infty} z^{(i)}(t) = 0. \tag{4.2.15}$$

Integrating (4.2.14) from t to ∞ repeatedly n times, we find that

$$\sum_{i=0}^{N} qp^i \int_{t}^{\infty} \frac{(t-s)^{n-1}}{(n-1)!} z(s-\sigma-i\tau)\,ds \leq z(t). \tag{4.2.16}$$

But by a result of Philos [131], if inequality (4.2.16) has an eventually positive solution $z(t)$, then the corresponding equation

$$\sum_{i=0}^{N} \int_{t}^{\infty} qp^i \frac{(t-s)^{n-1}}{(n-1)!} y(s - \sigma - i\tau) \, ds = y(t) \qquad (4.2.17)$$

also has an eventually positive solution $y(t)$. It follows then that equation (4.2.3) has the eventually positive solution $y(t)$. This contradicts condition H4.2.1 and completes the proof of the theorem.

Remark 4.2.1   For $N = 0$, condition H4.2.1 is equivalent to the condition

$$q^{1/n} \frac{\sigma}{n} > \frac{1}{e}. \qquad (4.2.18)$$

Indeed, when $N = 0$, equation (4.2.3) reduces to

$$y^{(n)}(t) + qy(t - \sigma) = 0 \qquad (4.2.19)$$

and (4.2.18) is a necessary and sufficient condition for all solutions of equation (4.2.19) to oscillate. See Ladas and Stavroulakis [106].

Thus we have the following corollary of Theorem 4.2.1.

Corollary 4.2.1   *Assume conditions (4.2.2) and (4.2.18) fulfilled and suppose that n is an odd natural number. Then every solution of equation (4.2.1) oscillates.*

The above corollary was established by Ladas and Sficas [100].

Remark 4.2.2   Assume that $n = 1$ and that

$$\sum_{i=0}^{\infty} qp^i(\sigma + i\tau) > \frac{1}{e}. \qquad (4.2.20)$$

Then there exists N large enough such that

$$\sum_{i=0}^{N} qp^i(\sigma + i\tau) > \frac{1}{e}. \qquad (4.2.21)$$

But (4.2.21) is a sufficient condition for all solutions of equation (4.2.3) to oscillate, see Hunt and Yorke [71]. Thus condition (4.2.20) implies that condition H4.2.1 is satisfied. The following corollary of Theorem 4.2.1 is established.

Corollary 4.2.2   *Assume conditions (4.2.2) and (4.2.20) are satisfied. Then every solution of the neutral delay differential equation*

$$\frac{\mathrm{d}}{\mathrm{d}t}[x(t) - px(t - \tau)] + qx(t - \sigma) = 0$$

oscillates.

The above corollary was proved by Györi [64].

Remark 4.2.3   In addition to condition (4.2.20), each of the following two conditions which were established in [106] implies that condition H4.2.1 is satisfied:

(i) $\displaystyle\max_{0 \leq i \leq N} \left[\frac{\sigma + i\tau}{n}(qp^i)^{1/n}\right] > \frac{1}{e}$;

(ii) $\displaystyle\left[q\frac{p^{N+1} - 1}{p - 1}\right]^{1/n}\frac{\sigma}{n} > \frac{1}{e}$ and $0 \leq p < 1$.

4.3   *n*th order nonlinear differential equations

This section sets down some oscillation criteria for the solutions of functional differential equations of the type

$$x^{(n)}(t) + \lambda x^{(n)}(t - \tau) + p(t)f(x(t - \tau)) = 0, \quad n \geq 1 \qquad (4.3.1)$$

where $\tau > 0$ is a constant delay and $\lambda > 0$ is an arbitrary constant. An analogous result for ordinary differential equations without delay was obtained by Kiguradze [80], and for equations with a retarded argument by Shevelo and Vareh [142].

Assume the following conditions (H4.3) fulfilled:

H4.3.1   The function $f(u) \colon \mathbb{R} \to \mathbb{R}$ is continuous, $uf(u) > 0$ for $u \neq 0$ and $\lim_{|u| \to +\infty} \inf |f(u)| > 0$.

H4.3.2   The function $p(t) \colon D \to [0, \infty)$ is continuous, where $D = [t_0 - \tau, +\infty)$, $t_0 \in \mathbb{R}$.

The alternative A is said to hold for equation (4.3.1) if for n even all of its solutions oscillate, while for n odd, they either oscillate or tend to zero for $t \to \infty$.

Define the operator L by the equality

$$(L\psi)(t) = \psi(t) + \lambda\psi(t - \tau) \qquad (4.3.2)$$

and denote by $\tilde{C}^k$ the space of functions $\psi(t): D \to \mathbb{R}$ locally having absolutely continuous derivatives up to kth order.

Lemma 4.3.1 ([81], p 243)

Let the following conditions hold:

(1) The function $\psi(t) \in \tilde{C}^{n-1}$ has a constant sign together with its derivatives up to nth order in the interval $[t_0, +\infty)$.

(2) For each $t \geq t_0$, the following inequality is valid

$$\psi(t)\psi^{(n)}(t) \leq 0 \quad (\psi(t)\psi^{(n)}(t) \geq 0).$$

Then there exists an integer $l, 0 \leq l \leq n$, such that $l + n$ is odd (even) and for $t \geq t_0$ the following inequalities hold

$$\psi(t)\psi^{(i)}(t) \geq 0, \quad i = 0, \ldots, l,$$

$$(-1)^{l+i}\psi(t)\psi^{(i)}(t) \geq 0, \quad i = l+1, \ldots, n;$$

$$|\psi^{(l-i)}(t)| \leq \frac{i!}{j!}(t - t_0)^{i-j}|\psi^{(l-j)}(t)|, \quad j = 0, \ldots, l; \quad i = 0, \ldots, j.$$

Moreover, if $l \neq 0$, then

$$|\psi(t)| \geq \sum_{i=l+1}^{n} \frac{1}{l!(i-l)!}(t - t_0)^{i-1}|\psi^{(i-1)}(t)|.$$

Theorem 4.3.1

Let the following conditions be fulfilled:

(1) Conditions (H4.3) hold.

(2) For any function $\psi(t) \in \tilde{C}^{n-1}$ such that $|\psi(t)| > 0$ for sufficiently large values of t and $\lim_{t \to \infty} \inf |(L\psi)(t)| > 0$, the inequality

$$\lim_{t \to +\infty} \inf |f(\psi(t - \tau))((L\psi)(t))^{-1}| > 0$$

is valid.

(3) There exists an absolutely continuous and nondecreasing function $\varphi(t): D \to (0, +\infty)$ such that for any measurable and closed set E having the property meas$(E \cap [t, t + 2\tau]) \geq \tau, t \in D$, *the following relations hold:*

$$\int_E (t - \tau)^{n-1} p(t)(\varphi(t - \tau))^{-1} \, dt = \infty \qquad (4.3.3)$$

$$\int_{t_0}^{\infty} \frac{dt}{t\varphi(t)} < \infty. \qquad (4.3.4)$$

Then the alternative A holds for equation (4.3.1).

Proof

Let $x(t)$ be a nonoscillating solution of equation (4.3.1), let the operator L be defined by equation (4.3.2) and suppose, for the sake of definiteness, that $x(t) > 0$ for $t \geq \bar{t}, \bar{t} \in D$.

Let the number n be even. Then equation (4.3.1) implies that for $t \geq \bar{t}$, $[(Lx)(t)]^{(n)} \leq 0$ and by virtue of Lemma 4.3.1 there exists a point $t_1 \geq \bar{t}$ and an integer l, $1 \leq l \leq n - 1$ such that for $t \geq t_1$ the following inequalities hold:

$$(Lx)(t)[(Lx)(t)]^{(i)} \geq 0, \quad i = 0, \ldots, l; \qquad (4.3.5)$$

$$(-1)^{l+i}(Lx)(t)[(Lx)(t)]^{(i)} \geq 0, \quad i = l + 1, \ldots, n; \qquad (4.3.6)$$

$$[(Lx)(t)]^{(l-i)} \leq \frac{i!}{j!}(t - t_1)^{i-j}[(Lx)(t)]^{(l-j)}. \qquad (4.3.7)$$

Multiplying both sides of equation (4.3.1) by the function

$$\frac{(t - \tau)^{n-l}}{\varphi(t)[(Lx)(t)]^{(l-1)}}$$

and taking into account that $t - \tau > 0$, we obtain

$$\frac{t^{n-l}[(Lx)(t)]^{(n)}}{\varphi(t)[(Lx)(t)]^{(l-1)}} + \frac{p(t)f(x(t-\tau))(t-\tau)^{n-l}}{\varphi(t)[(Lx)(t)]^{(l-1)}} \leq 0. \qquad (4.3.8)$$

Inequality (4.3.7) for $i = 1, j = l$ yields the following inequality for $t \geq t_1$:

$$(t - t_1)^{l-1}[(Lx)(t)]^{(l-1)} \leq l!(Lx)(t). \qquad (4.3.9)$$

On the other hand, there exists a point $t_2 \geq t_1$ such that for $t \geq t_2$, $t - \tau \geq 2t_1$ holds. Hence, if $t \geq t_2$, inequalities (4.3.8) and (4.3.9) imply the inequality

$$\frac{t^{n-l}[(Lx)(t)]^{(n)}}{\varphi(t)[(Lx)(t)]^{(l-1)}} + \frac{cp(t)f(x(t-\tau))(t-\tau)^{n-1}}{\varphi(t)(Lx)(t)} \leq 0 \qquad (4.3.10)$$

where $c > 0$ is a constant.

Since $[(Lx)(t)] \geq 0$ and $x(t)$ is a regular solution, there exists a point $t_3 \geq t_2$ such that $(Lx)(t) \geq c_1 > 0$ for $t \geq t_3$ and in virtue of Lemma 1 from [160], there exists a closed and measurable set E with the property $\text{meas}(E \cap [t, t + 2\tau]) \geq \tau$ for $t \geq t_3$ such that $x(t - \tau) \geq c_2 > 0$ for all $t \in E$. Integrating inequality (4.3.10) over the set $E \cap [t_3, t], t > t_3$ and taking into account that for $t \geq t_3$, $[(Lx)(t)]^{(n)} \leq 0$ and $[(Lx)(t)]^{(l-1)} \geq 0$, we obtain the inequality

$$\int_{t_3}^{t} \frac{s^{n-l}[(Lx)(s)]^{(n)} \, ds}{\varphi(s)[(Lx)(s)]^{(l-1)}} + \int_{E \cap [t_3, t]} \frac{cp(s)f(x(s-\tau))(s-\tau)^{n-1} \, ds}{\varphi(s)(Lx)(s)} \leq 0.$$

$$(4.3.11)$$

Integrating by parts the first integral in (4.3.11), we obtain

$$\frac{\sum_{i=0}^{n-l+1} (-1)^i \dfrac{(n-l)!}{(i+1)!} s^{i+1}[(Lx)(s)]^{(l+i)}}{\varphi(s)[(Lx)(s)]^{(l-1)}} \Bigg|_{t_3}^{t}$$

$$- \int_{t_3}^{t} \sum_{i=0}^{n-l+1} (-1)^i s^{i+1} \frac{(n-l)!}{(i+1)!} [(Lx)(s)]^{(l+i)} \, d[(\varphi(s)[(Lx)(s)]^{(l-1)})^{-1}]$$

$$+ \int_{E \cap [t_3, t]} \frac{cp(s)f(x(s-\tau))(s-\tau)^{n-1} \, ds}{\varphi(s)[(Lx)(s)]}$$

$$- (n-l)! \int_{t_3}^{t} \frac{[(Lx)(s)]^{(l)} \, ds}{\varphi(s)[(Lx)(s)]^{(l-1)}} \leq 0. \qquad (4.3.12)$$

From inequalities (4.3.5) and (4.3.6) and condition (2) of Theorem 4.3.1 we can conclude that for $t \geq t_3$ all derivatives $[(Lx)(t)]^{(i)}$ of an even order will be nonnegative and monotone decreasing, while all derivatives of an odd order will be nonpositive and monotone increasing, hence the sum participating in inequality (4.3.12) is nonnegative.

For $t \geq t_3$ inequality (4.3.5) and condition (3) of Theorem 4.3.1 yield

$$d[(\varphi(t)[(Lx)(t)]^{(l-1)})^{-1}] \leq 0.$$

Hence we can conclude that for $t \geq t_3$ the first two addends in the right-hand side of inequality (4.3.12) are nonnegative and therefore the following inequality holds:

$$\int_{E \cap [t_3, t]} \frac{cp(s)f(x(s-\tau))(s-\tau)^{n-1}\, ds}{\varphi(s)(Lx)(s)} \leq (n-1)! \int_{t_3}^{t} \frac{[(Lx)(s)]^{(l)}\, ds}{\varphi(s)[(Lx)(s)]^{(l-1)}}.$$

$$(4.3.13)$$

On the other hand, (4.3.7) implies that for $t \geq t_3, j = 1, i = 0$, the inequality

$$[(Lx)(t)]^{(l)}(t - t_3) \leq [(Lx)(t)]^{(l-1)}$$

holds. Furthermore, there exists a point $t_4 \geq t_3$, such that for $t \geq t_4$ the inequality $t - t_3 \geq \frac{1}{2}(t - \tau)$ holds and hence

$$\frac{1}{2}[(Lx)(t)]^{(l)}(t - \tau) \leq [(Lx)(t)]^{(l-1)}. \qquad (4.3.14)$$

Since condition (2) of Theorem 4.3.1 implies that there exists a point $t_5 \geq t_4$ and a constant $c_3 > 0$ such that for $t \geq t_5$ we have

$$f(x(t - \tau))((Lx)(t))^{-1} \geq c_3/2$$

then inequalities (4.3.13) and (4.3.14) yield the inequality

$$\frac{cc_3}{2} \int_{E \cap [t_5, t]} \frac{p(s)(s-\tau)^{n-1}\, ds}{\varphi(s)} \leq 2(n-1)! \int_{t_5}^{t} \frac{ds}{(s-\tau)\varphi(s)}.$$

Passing to the limit in the above inequality as $t \to \infty$ and taking into account inequality (4.3.4), we get

$$\frac{cc_3}{2} \int_{E \cap [t_5, \infty)} \frac{(t-\tau)^{n-1}p(t)}{\varphi(t)}\, dt \leq 2(n-1)! \int_{t_5}^{\infty} \frac{dt}{(t-\tau)\varphi(t)} < \infty$$

which contradicts equality (4.3.3).

Let n be odd and suppose that the equation has a nonoscillating solution $x(t)$. Without loss of generality we can assume that $x(t) > 0$ for $t \geq \bar{t}$, $\bar{t} \in D$. Then, since for $t \geq \bar{t}$

$$(Lx)(t) \geq 0, \quad [(Lx)(t)]^{(n)} \leq 0$$

Lemma 4.3.1 implies that there exists a point $t_1 \geq \bar{t}$ and an integer l, $0 \leq l < n$, $l + n$ being odd, such that for $t \geq t_1$ inequalities (4.3.5) and (4.3.6) hold, while if $l \neq 0$, inequality (4.3.7) holds.

If $l \geq 2$, then the arguments are as in the case when n is even. In the case $l = 0$ and $\lim_{t \to \infty} (Lx)(t) = 0$ we have $\lim_{t \to \infty} x(t) = 0$.

Let $l = 0$ and $\lim_{t \to \infty} (Lx)(t) = c_4 > 0$.

Since $\lim_{t \to \infty} (Lx)(t) = c_4$, there exists a point $t_2 \geq t_1$ such that $(Lx)(t) \geq c_4/2$ for $t \geq t_2$. In virtue of Lemma 1 from [160] there exists a measurable and closed set E, $E \subset [t_2, \infty)$ with the property

$$\text{meas}(E \cap [t, t + 2\tau]) \geq \tau, \quad t \geq t_2$$

such that $x(t - \tau) \geq c_5 > 0$ for $t \in E$. Hence there exists a constant $c_6 > 0$ such that

$$\inf_{t \in E} f(x(t - \tau)) \geq c_6. \tag{4.3.15}$$

Multiplying equation (4.3.1) by t^{n-1} and integrating from t_2 to $t \geq t_2$, we obtain

$$\int_{t_2}^{t} s^{n-1}[(Lx)(s)]^{(n)} \, ds + \int_{t_2}^{t} s^{n-1} p(s) f(x(s - \tau)) \, ds = 0. \tag{4.3.16}$$

Integrating by parts in the first integral in (4.3.16) $n - 1$ times, we get

$$s^{n-1}[(Lx)(s)]^{(n-1)} \big|_{t_2}^{t} - (n-1) s^{n-2}[(Lx)(s)]^{(n-2)} \big|_{t_2}^{t}$$

$$+ \cdots + (n-1)!(Lx)(s)\big|_{t_2}^{t} + \int_{t_2}^{t} s^{n-1} p(s) f(x(s - \tau)) \, ds = 0. \tag{4.3.17}$$

Since $[(Lx)(t)]' \leq 0$, inequality (4.3.6) implies that all derivatives of $(Lx)(t)$ of even order are nonnegative, hence inequalities (4.3.15) and (4.3.17) yield the inequality

$$c_6 \int_{E \cap [t_2, t]} s^{n-1} p(s) \, ds \leq t_2^{n-1}[(Lx)(t_2)]^{(n-1)} - (n-1) t_2^{n-2}[(Lx)(t_2)]^{(n-2)}$$

$$+ \cdots + (n-1)!\,(Lx)(t_2).$$

Passing to the limit in the last inequality as $t \to \infty$, we obtain the inequality

$$\int_{E \cap [t_2, \infty)} t^{n-1} p(t) \, dt < \infty. \tag{4.3.18}$$

Taking into account that $t - \tau < t$ and $\varphi(t)$ is a nondecreasing function, (4.3.18) yields

$$\int_{E \cap [t_2, \infty)} (t - \tau)^{n-1} p(t) (\varphi(t - \tau))^{-1} \, dt < \infty$$

which contradicts equality (4.3.3).

Thus Theorem 4.3.1 is proved.

Since condition (2) of Theorem 4.3.1 is difficult to verify, we proceed to prove, by means of an indirect criterion, the validity of the alternative A for equation (4.3.1) in the case when $f(u)$ is differentiable.

Theorem 4.3.2

Let the following conditions be fulfilled:

(1) Conditions ($H4.3$) hold.

(2) The function $f \in C^1(\mathbb{R}; \mathbb{R})$ and $f'(u) \geq 0$, $u \in \mathbb{R}$.

(3) There exists a function $\varphi \in C^1(D; \mathbb{R})$, $\varphi(t) > 0$, $\varphi'(t) \geq 0$ for $t \in D$, such that for each closed measurable set $E \subset D$ with the property $\mathrm{meas}(E \cap [t, t + 2\tau]) \geq \tau$, $t \in D$, the following relations hold

$$\int_E \frac{(t - \tau)^{n-1}}{\varphi(t)} \, dt = +\infty \tag{4.3.19}$$

$$\int_\varepsilon^{+\infty} \frac{du}{f(u)\varphi(u^{-(n-1)})} < \infty, \quad \int_{-\varepsilon}^{-\infty} \frac{du}{f(u)\varphi((-u)^{-(n-1)})} < \infty,$$

$$\varepsilon > 0. \tag{4.3.20}$$

Then the alternative A holds for equation (4.3.1).

Proof

Let $x(t)$ be a nonoscillating solution of equation (4.3.1), for the sake of definiteness suppose that $x(t) > 0$ for $t \geq \bar{t}$, $\bar{t} \in D$, and let the operator L be defined by equality (4.3.2). Then Lemma 4.3.1 implies that there exists a point $t_1 \geq \bar{t}$ and a number l, $0 \leq l \leq n$, $l + n$ odd, such that for $t \geq \bar{t}$ inequalities (4.3.5) and (4.3.6) hold, while if $l \neq 0$, then inequality (4.3.7) also holds.

Let n be an even number. Then, since $l \geq 1$ by virtue of Lemma 1 from [160], there exists a set $E \subset D$ such that $\text{meas}(E \cap [t, t + 2\tau]) \geq \tau$, $t \in D$, and $x(t - \tau) \geq C_7 > 0$ for $t \in E$. Besides, $l \geq 1$ and (4.3.7) implies that for $t \geq t_1$ the following inequality holds:

$$[(Lx)(t)]^{(l)} \leq j!(t - t_1)^{-j}[(Lx)(t)]^{(l-j)}, \quad j = 0, \ldots, l.$$

If we choose a point $t_2 \geq t_1$ such that for $t \geq t_2$ we have $t - \tau \geq 2t_1$ then from the last inequality it follows that

$$[(Lx)(t)]^{(l)} \leq 2^{l-1}(l-1)!(t - \tau)^{1-l}[(Lx)(t)], \quad t \geq t_2 \quad (4.3.21)$$

There exists also a point $t_3 \geq t_2$ and a constant C_8 such that

$$\inf_{t \in E \cap [t_3, \infty)} f(x(t - \tau))[f((Lx)(t))]^{-1} \geq C_8. \quad (4.3.22)$$

Multiply equation (4.3.1) by the function $(t - \tau)^{n-1}[\varphi(t)f((Lx)(t))]^{-1}$ and integrate from t_3 to $t > t_3$. We get

$$\int_{t_3}^{t} \frac{p(s)(s - \tau)^{n-1}f(x(s - \tau))\, ds}{\varphi(s)f((Lx)(s))} = \int_{t_3}^{t} \frac{(s - \tau)^{n-1}[(Lx)(t)]^{(n)}\, ds}{\varphi(s)f((Lx)(s))}$$

from which, integrating by parts in the integral in the right-hand side and taking into account equality (4.3.22), we have

$$C_8 \int_{E \cap [t_3, t]} \frac{(s - \tau)^{n-1}p(s)\, ds}{\varphi(s)}$$

$$= -\frac{(s - \tau)^{n-1}[(Lx)(t)]^{(n-1)}}{\varphi(s)f((Lx)(s))}\Bigg|_{t_3}^{t}$$

$$+ (n - 1) \int_{t_3}^{t} \frac{(s - \tau)^{n-2}[(Lx)(s)]^{(n-1)}\, ds}{\varphi(s)f((Lx)(s))}$$

$$+ \int_{t_3}^{t} (s - \tau)^{n-1}[(Lx)(s)]^{(n-1)}\, d[(\varphi(s)f((Lx)(s)))^{-1}]. \quad (4.3.23)$$

Conditions (2) and (3) of Theorem 4.3.2 and (4.3.6) yield that

$$[(Lx)(t)]^{(n-1)} \geq 0, \quad d[(\varphi(t)f((Lx)(t)))^{-1}] \leq 0$$

and hence from (4.3.23) we obtain the inequality

$$C_8 \int_{E \cap [t_3, t]} \frac{(s-\tau)^{n-1}p(s)}{\varphi(s)} ds \leq C_9 + (n-1) \int_{t_3}^{t} \frac{(s-\tau)^{n-1}[(Lx)(s)]^{(n-1)} ds}{\varphi(s)f((Lx)(s))}$$

$$(4.3.24)$$

where $C_9 > 0$ is a constant. Integrating by parts the right-hand side of (4.3.24) $n-l$ times, we obtain

$$C_8 \int_{E \cap [t_3, t]} \frac{(s-\tau)^{n-1}p(s)}{\varphi(s)} ds$$

$$\leq C_9 + (n-1) \frac{\sum_{i=l}^{n-2} (-1)^i \dfrac{(n-2)!}{i!} (s-\tau)^i [(Lx)(s)]^{(i)}}{\varphi(s)f((Lx)(s))} \Bigg|_{t_3}^{t}$$

$$- (n-1) \int_{t_3}^{t} \sum_{i=l}^{n-2} (-1)^i \frac{(n-2)!}{i!} (s-\tau)^i$$

$$\times [(Lx)(s)]^{(i)} d[(\varphi(s)f((Lx)(s)))^{-1}]$$

$$+ (-1)^{n-l-1} \frac{(n-1)!}{(l-1)!} \int_{t_3}^{t} \frac{(s-\tau)^{l-1}[(Lx)(s)]^{(l)} ds}{\varphi(s)f((Lx)(s))}. \qquad (4.3.25)$$

Since (4.3.5) and (4.3.6) imply for $s \geq t_3$ the inequality

$$\sum_{i=l}^{n-2} (-1)^i \frac{(n-2)!}{i!} (s-\tau)^i [(Lx)(s)]^{(i)} \leq 0$$

holds, then (4.3.21) and (4.3.25) yield

$$C_8 \int_{E \cap [t_3, t]} \frac{(s-\tau)^{n-1}p(s)}{\varphi(s)} ds \leq C_{10} + 2^{l-1}(n-1)! \int_{t_3}^{t} \frac{[(Lx)(s)]'}{\varphi(s)f((Lx)(s))} ds.$$

$$(4.3.26)$$

Taylor's theorem and the fact that $[(Lx)(t)]^{(n)} \leq 0$ for $t \geq t_3$ imply that there exists a constant $a \geq 1$ such that $(Lx)(t) \leq at^{n-1}$ for $t \geq t_3$. Then conditions (2) and (3) of Theorem 4.3.2 and $(4.3.5)$ imply the inequality

$$\int_{t_3}^{t} \frac{[(Lx)(s)]'}{\varphi(s)f((Lx)(s))} ds \leq \int_{t_3}^{t} \frac{d[(Lx)(s)]}{f((Lx)(s))\varphi\left(\left[\frac{(Lx)(s)}{a}\right]^{-(n-1)}\right)}$$

$$= \int_{a^{-1}(Lx)(t_3)}^{a^{-1}(Lx)(t)} \frac{du}{f(au)\varphi(u^{-(n-1)})}.$$

The last inequality and inequalities $(4.3.20)$ and $(4.3.26)$ yield the inequality

$$\int_{E \cap [t_3, \infty)} \frac{(t-\tau)^{n-1}p(t)}{\varphi(t)} dt \leq a \int_{a^{-1}(Lx)(t_3)}^{\infty} \frac{du}{[f(u)\varphi(u^{-(n-1)})} < \infty$$

which contradicts condition $(4.3.19)$.

Let n be odd and let the nonoscillating solution $x(t)$ of $(4.3.1)$ and the operator L satisfy the same assumptions as in the case when the number n is even. If for the numbers l, $0 \leq l \leq n$, which exist in view of Lemma 4.3.1, the condition $l \geq 2$ holds, then by arguments analogous to those in the case when n is even we get to a contradiction. Therefore, $l = 0$ (n odd) and since $(4.3.6)$ implies that $[(Lx)(t)]' \leq 0$, then either $\lim_{t \to \infty}(Lx)(t) = 0$, and hence $\lim_{t \to \infty} x(t) = 0$, or $\lim_{t \to \infty}(Lx)(t) = C_{11} > 0$.

Therefore, by Lemma 1 from [160], there exists a closed and measurable set $E \subset D$, $\text{meas}(E \cap [t, t + 2\tau]) \geq \tau$, $t \geq \bar{t}$, such that $x(t - \tau) \geq C_{12} > 0$ for $t \in E$. Multiplying equation $(4.3.1)$ by t^{n-1} and integrating from $\bar{t}$ to $t > \bar{t}$, we obtain

$$\sum_{i=1}^{n} (-1)^{i+1} \frac{(n-1)!}{(n-i)!} s^{n-i}[(Lx)(s)]^{(n-i)}|_{\bar{t}}^{t} + C_{13} \int_{E \cap [\bar{t}, t]} s^{n-1}p(s)\, ds \leq 0,$$

$$C_{13} > 0.$$

Since $(4.3.6)$ implies that for $t \geq \bar{t}$ all derivatives of $(Lx)(t)$ of even order are nonpositive and monotone increasing, while those of odd order are nonnegative and monotone decreasing, then the last inequality, passing to the limit as $t \to \infty$, yields the inequality

$$\int_{E \cap [\bar{t}, \infty)} t^{n-1} p(t) \, dt < \infty$$

from which, since $\varphi'(t) \geq 0$, $\varphi(t) > 0$, we obtain the inequality

$$\int_{E \cap [\bar{t}, \infty)} [t^{n-1} p(t)(\varphi(t))^{-1}] \, dt < \infty$$

which contradicts (4.3.19).
     Thus Theorem 4.3.2 is proved.

Remark 4.3.1   For $n = 1$ the proofs of Theorems 4.3.1 and 4.3.2 can be considerably simplified since the integration by parts is omitted.
     We shall show by a counter-example that equality (4.3.19) from condition (3) of Theorem 4.3.2 cannot be replaced by the weaker classical condition

$$\int_{t_0}^{\infty} [(t - \tau)^{n-1} p(t)(\varphi(t))^{-1}] \, dt = \infty. \tag{4.3.27}$$

     Consider the equation

$$x'(t + \pi) + x'(t) + p(t) x^3(t) = 0 \tag{4.3.28}$$

where $t \geq t_0 > 0$ and $p(t) = [t^2 + (t + \pi)^2][t^2(t + \pi)^2(t^{-1} + 1 - \cos t)^3]^{-1}$. Here $f(u) = u^3$, and let $\varphi(t) \equiv 1$. After simple calculations we obtain

$$\int_{t_0}^{\infty} p(t) \, dt$$

$$\geq \sum_{k = [t_0] + 1}^{\infty} \int_{2k\pi - k^{-1}}^{2k\pi + k^{-1}} (t^2 + (t + \pi)^2) t^{-2} (t + \pi)^{-2} (t^{-1} + 1 - \cos t)^{-3} \, dt$$

$$\geq \sum_{k = [t_0] + 1}^{\infty} 4k^{-1} (2k\pi + k^{-1} + \pi)^{-2} ([2k\pi - k^{-1}]^{-1} + 1 - \cos k^{-1})^{-3}$$

which yields

$$\int_{t_0}^{\infty} p(t)\, dt = \infty.$$

On the other hand, if

$$E = \bigcup_{k = [t_0] + 1}^{\infty} \left\{ t : t \geq t_0, \ \frac{\pi}{4} + 2k\pi \leq t \leq \frac{5\pi}{4} + 2k\pi \right\}$$

then

$$\int_{E} p(t)\, dt \leq \int_{E} \frac{[t^2 + (t + \pi)^2]}{t^2 (t + \pi)^2 \left(t^{-1} + 1 - \dfrac{1}{\sqrt{2}} \right)^3}\, dt < \infty$$

which shows that $p(t)$ satisfies the classical condition (4.3.27) but does not satisfy (4.3.19).

It can be easily verified that equation (4.3.28) has a solution $x(t) = t^{-1} + 1 - \cos t$. Thus condition (4.3.19) is essential.

4.4   Notes and comments to Chapter 4

The results of Section 4.1 are due to Grammatikopoulos, Ladas and Meimaridou [48]. Oscillatory properties of the solutions of linear differential equations of neutral type were investigated by Jaroš and Kusano [77], [78].

The results of Section 4.2 are due to Grove, Ladas and Schinas [58]. Criteria for oscillation and asymptotic behaviour of the solutions of neutral differential equations with constant coefficients were obtained by Ladas and Sficas [100].

The results of Section 4.3 are due to Zahariev and Bainov [159]. Sufficient conditions for the existence of bounded nonoscillating solutions of functional differential equations of neutral type were obtained by Bainov et al [13]. Some general approaches to the investigation of the oscillatory and asymptotic properties of neutral equations were given by Myshkis et al [125], Jaroš and Kusano [75] and Drakhlin [25]. Oscillatory properties of the solutions of neutral integro-differential equations were investigated by Bainov et al [10] and Ruan [137]. Estimates of the interval of nonoscillation for equations of neutral type were obtained by Domoshnitskii [18], [19], [20]. Sufficient conditions for oscillation of the solutions of some classes of nth order neutral equations were obtained by Grace and Lalli [42].

5

Systems of ordinary differential equations of neutral type

5.1 Linear systems of ordinary differential equations with constant coefficients

In the oscillation theory of linear delay differential equations one of the most important objectives is to give a necessary and sufficient condition for oscillation via the characteristic equation. Such a result for scalar delay differential equations was proved by Arino *et al* [6], Ladas and Stavroulakis [105] and Tramov [148] using various methods, and the method of [6] was extended to delay systems in [34]. In the neutral case many special scalar equations have been investigated using different techniques, but at this time a general result is not known.

In this section it is proved that in the general scalar as well as system cases a neutral delay differential equation with several delays has a non-oscillating solution if and only if its characteristic equation has a real root. The proof is elementary and is based on the method of the Laplace transform, using a fact from [66] that the solutions of a neutral equation do not grow faster than exponentially.

Consider the neutral delay differential equation of the form

$$\frac{\mathrm{d}}{\mathrm{d}t}\left[x(t) - \sum_{j=1}^{m} B_j x(t - \sigma_j) \right] = \sum_{i=1}^{k} A_i x(t - \tau_i) \qquad (5.1.1)$$

where we assume the following conditions fulfilled:

H5.1.1 A_i $(1 \leq i \leq k)$ and B_j $(1 \leq j \leq m)$ are given $n \times n$-matrices, $\tau_i (1 \leq i \leq k)$ and $\sigma_j (1 \leq j \leq m)$ are given positive constants, $\gamma = \max\{\tau, \sigma\}$, $\tau = \max_{1 \leq i \leq k} \tau_i$, $\sigma = \max_{1 \leq j \leq m} \sigma_j$.

Definition 5.1.1   By a solution of (5.1.1) on the interval $[-\gamma, \infty)$ we mean a function $x \in C([-\gamma, \infty); \mathbb{R}^n)$ such that $x(t) - \sum_{j=1}^{m} B_j x(t - \sigma_j)$ is continuously differentiable and satisfies (5.1.1) on $[0, \infty)$.

Definition 5.1.2   A solution $x = (x_1, \ldots, x_n)^T : [-\gamma, \infty) \to \mathbb{R}^n$ of (1.5.1) is said to be nonoscillating if there exist $t_0 \geq 0$ and $i_0 \in \{1, \ldots, n\}$ such that $|x_{i_0}(t)| > 0$, $t \geq t_0$.

Remark 5.1.1   If equation (5.1.1) has a nonoscillating solution, then there exists a solution $x = (x_1, \ldots, x_n)^T : [-\gamma, \infty) \to \mathbb{R}^n$ of (5.1.1) such that for some $i_0 \in \{1, \ldots, n\}$, $x_{i_0}(t) > 0$, $t \geq -\gamma$. In fact, if $y(t)$ is a nonoscillating solution of (5.1.1), then there exists an index $i_0 \in \{1, \ldots, n\}$ and $t_0 \geq 0$ such that $|y_{i_0}(t)| > 0$, $t \geq t_0$. But it is easily seen that $x(t) = (\text{sign } y_{i_0}(t))y(t + \gamma + t_0)$ is a solution of (5.1.1) and $x_{i_0}(t) = |y_{i_0}(t)| > 0$, $t \geq -\gamma$.

Theorem 5.1.1

Assume H5.1.1 fulfilled. Then equation (5.1.1) has a nonoscillating solution if and only if its characteristic equation

$$\det\left(\lambda I - \lambda \sum_{j=1}^{m} B_j \, e^{-\lambda \sigma_j} - \sum_{i=1}^{n} A_i \, e^{-\lambda \tau_i} \right) = 0 \qquad (5.1.2)$$

has a real root.

Proof

If equation (5.1.2) has a real root λ then there exists a constant vector $c = (c_1, \ldots, c_n)^T$ such that $\sum_{i=1}^{n} |c_i| > 0$ and

$$x(t) = e^{\lambda t} c \qquad (5.1.3)$$

is a solution of (5.1.1) on $(-\infty, \infty)$. But it is easily seen that $x(t)$ is a nonoscillating solution of (5.1.1).

Therefore the proof will be complete if we show that if equation (5.1.1) has a nonoscillating solution on $[-\gamma, \infty)$, then equation (5.1.2) has a real root. Suppose that this is not true, that is, (5.1.1) has a nonoscillating solution on $[-\gamma, \infty)$ and (5.1.2) has no real root. We show that it leads to a contradiction.

By Remark 5.1.1 we have that if (5.1.1) has a nonoscillating solution, then it has a solution $x(t) = (x_1(t), \ldots, x_n(t))^T$ such that for an index $i_0 \in \{1, \ldots, n\}$

$$x_{i_0}(t) > 0, \quad t \geq -\gamma. \tag{5.1.4}$$

But from [66] we know that the solutions of (5.1.1) do not grow faster than exponentially, therefore

$$|x_i(t)| \leq a\, e^{bt}, \quad t \geq -\gamma, \quad 1 \leq i \leq n \tag{5.1.5}$$

for some constants $a > 0$ and $b \in \mathbb{R}$.

Thus the Laplace transform $L[x]$ defined by

$$L[x](s) = \int_0^\infty e^{-st} x(t)\, dt \tag{5.1.6}$$

exists and is an analytic function of s for Re $s > b$. But (5.1.5) implies that the function

$$u(t) = x(t) - \sum_{j=1}^m B_j x(t - \sigma_j), \quad t \geq 0$$

and its derivative $u'(t)$ have the same exponential growth b as $x(t)$. Thus

$$L[u'](s) = sL[u](s) - u(0)$$

exists for Re $s > b$, and from (5.1.1) it follows that

$$sL[x](s) - s\sum_{j=1}^m B_j e^{-s\sigma_j} L[x](s) - x(0) + \sum_{j=1}^m B_j x(t - \sigma_j)$$

$$- s\sum_{j=1}^m B_j e^{-s\sigma_j} \int_{-\sigma_j}^0 x(t) e^{-st}\, dt$$

$$= \sum_{i=1}^k A_i e^{-s\tau_i} L[x](s) + \sum_{i=1}^k A_i e^{-s\tau_i} \int_{-\tau_i}^0 x(t) e^{-st}\, dt, \quad \text{Re } s > b \tag{5.1.7}$$

Here, we used the simple fact that

$$L[x(t-k)](s) = \int_0^\infty e^{-st} x(t-k)\,dt$$

$$= e^{-sk} L[x](s) + e^{-sk} \int_{-k}^{0} x(t)\,e^{-st}\,dt, \quad \operatorname{Re} s > b.$$

for any constant $k \in [0, \gamma]$.

From (5.1.7) we have

$$\Delta(s)L[x](s) = \varphi(s), \quad \operatorname{Re} s > b \tag{5.1.8}$$

where

$$\Delta(s) = sI - s \sum_{j=1}^{m} B_j e^{-s\sigma_j} - \sum_{i=1}^{k} A_i e^{-s\tau_i} \tag{5.1.9}$$

and

$$\varphi(s) = x(0) - \sum_{j=1}^{m} B_j x(t - \sigma_j) + s \sum_{j=1}^{m} B_j e^{-s\sigma_j} \int_{-\sigma_j}^{0} x(u)\,e^{-su}\,du$$

$$+ \sum_{i=1}^{k} A_i e^{-s\tau_i} \int_{-\tau_i}^{0} x(u)\,e^{-su}\,du \tag{5.1.10}$$

are analytic functions of s for all s.

Since (5.1.2) has no real root, we have

$$a(s) = \det \Delta(s) \neq 0, \quad s \in \mathbb{R}$$

therefore (5.1.8) implies

$$L[x](s) = \Delta^{-1}(s)\varphi(s) \tag{5.1.11}$$

for any $s \in [b, \infty)$, where the right-hand side is analytic on $\mathbb{R}$. Denote by $\psi_{i_0}(s)$ the i_0th component of $\Delta^{-1}(s)\varphi(s)$, that is $\psi_{i_0}(s) = (\Delta^{-1}(s)\varphi(s))_{i_0}$.

Then (5.1.11) yields

$$L[x_{i_0}](s) = \psi_{i_0}(s) \tag{5.1.12}$$

for $\operatorname{Re} s > b$.

Formula (5.1.12) shows that $L[x_{i_0}]$ has an analytic extension on the whole real axis. It will be shown that this extension can in fact be expressed as a Laplace transform, that is, for every s in $\mathbb{R}$, $e^{-st} x_{i_0}(t)$ is in $L_1(\mathbb{R}^+)$.

First observe that the set

$$\{s \in \mathbb{R} : e^{-st} x_{i_0}(t) \in L_1(\mathbb{R}^+)\}$$

is an interval (s_0, ∞). On this interval we have

$$\int_0^\infty e^{-st} x_{i_0}(t)\,dt = \psi_{i_0}(s)$$

and because ψ_{i_0} is continuous and $x_{i_0} \geq 0$, it follows from Beppo–Levi's theorem that the integral is also finite for $s = s_0$ (if $s_0 > -\infty$). Assume for the moment that $s_0 > -\infty$. We choose a point $s_1 > s_0$ such that the Taylor series associated with ψ_{i_0} at $s = s_1$ has a radius of convergence exceeding $s_1 - s_0$. We can express the coefficients of the Taylor series of ψ_{i_0} at s_1 in terms of x_{i_0}:

$$\frac{\psi_{i_0}^{(k)}(s_1)}{k!} = \frac{(-1)^k}{k!} \int_0^\infty t^k e^{-s_1 t} x_{i_0}(t)\,dt. \tag{5.1.13}$$

Inside the disc of convergence, the series

$$\sum_{k \geq 0} \frac{|\psi_{i_0}^{(k)}(s_1)|}{k!} (s_1 - s)^k$$

also converges. This implies that

$$\sum_{k \geq 0} \frac{1}{k!} (s_1 - s)^k \int_0^\infty t^k e^{s_1 t} x_{i_0}(t)\,dt < \infty \tag{5.1.14}$$

for s in some interval $(s_1 - \delta, s_0)$, for some $\delta > 0$.

But this expression can be regarded as a sum of integrals of positive functions. Beppo–Levi's theorem then applies and leads to the conclusion that

$$\int_0^\infty e^{-st} x_{i_0}(t)\, dt < \infty$$

for s in $(s_0 - \delta, s_0)$.

This contradicts the definition of s_0 and leads to the conclusion that formula (5.1.12) holds on the whole real axis.

Now consider the behaviour of $\psi_{i_0}(s)$ as $s \to -\infty$. From (5.1.10) it is easily seen that there exists $a_1 > 0$ and $a_2 > 0$ such that

$$|\varphi(s)| \leq a_1\, e^{a_2|s|}, \quad s \in \mathbb{R}. \tag{5.1.15}$$

One can see that for any $n \times n$-matrix A, $\det A > 0$ implies

$$|A^{-1}| \leq \frac{b}{\det A}\, |A|^{n-1}$$

where b is a constant depending only on n. Thus

$$|\Delta^{-1}(s)| \leq \frac{b}{a(s)}\, |\Delta(s)|^{n-1}, \quad s \in \mathbb{R} \tag{5.1.16}$$

where $a(s) = \det \Delta(s) > 0$. But (5.1.9) implies that for some constants $c_1 > 0$ and $c_2 > 0$,

$$|\Delta(s)| \leq c_1\, e^{c_2|s|}, \quad s \in \mathbb{R}. \tag{5.1.17}$$

Moreover, $a(s)$ is a polynomial of s, $e^{-s\sigma_j}$ $(1 \leq j \leq m)$ and $e^{-s\tau_i}$ $(1 \leq i \leq k)$, where $\sigma_j > 0$ $(1 \leq j \leq m)$ and $\tau_i > 0$ $(1 \leq i \leq k)$. On the other hand, $a(s) > 0$, $s \in \mathbb{R}$, therefore $\lim_{s \to -\infty} a(s) = +\infty$. Thus there exists $s_1 \in (-\infty, 0)$ such that

$$1/a(s) \leq 1, \quad s \leq s_1.$$

Thus (5.1.16) and (5.1.17) imply

$$|\Delta^{-1}(s)| \leq b(c_1)^{n-1}\, e^{(n-1)c_2|s|}, \quad s \leq s_1$$

and (5.1.15) yields

$$|\Delta^{-1}(s)\varphi(s)| \leq d_1\, e^{d_2|s|}, \quad s \leq s_1$$

where $d_1 > 0$ and $d_2 > 0$ are suitable constants. Since (5.1.12) holds for all s, it follows that

$$0 < L[x_{i_0}](s) = \psi_{i_0}(s) \leq |\Delta^{-1}(s)\varphi(s)| \leq d_1 e^{d_2|s|}$$

for any $s \leq s_1$, therefore

$$0 \leq \lim_{s \to -\infty} \sup e^{d_2|s|} L[x_{i_0}](s) \leq d_1. \tag{5.1.18}$$

But for $T > d_2$, we obtain

$$e^{-d_2|s|} L[x_{i_0}](s) = e^{-d_2|s|} \int_0^\infty e^{-st} x_{i_0}(t)\, dt$$

$$\geq e^{-d_2|s|} \int_T^\infty e^{-st} x_{i_0}(t)\, dt \geq e^{-d_2|s|} e^{T|s|} \int_T^\infty x_{i_0}(t)\, dt \to +\infty$$

as $s \to -\infty$, since $x_{i_0}(t) > 0$ $(t \geq 0)$. This contradicts (5.1.18), therefore equation (5.1.2) indeed has a real root. The proof of the theorem is complete.

In the scalar case $(n = 1)$ we have

Corollary 5.1.1   *Let b_j and $\sigma_j \geq 0$ $(1 \leq j \leq m)$ and a_i and $\tau_i \geq 0$ $(1 \leq i \leq k)$ be given real numbers. Then the equation*

$$\frac{d}{dt}\left[x(t) - \sum_{j=1}^m b_j x(t - \sigma_j) \right] = \sum_{i=1}^k a_i x(t - \tau_i)$$

is oscillatory if and only if its characteristic equation

$$\lambda - \lambda \sum_{j=1}^m b_j e^{-\lambda \sigma_j} = \sum_{i=1}^k a_i e^{-\lambda \tau_i}$$

has no real root.

Since any higher order equation is equivalent to a first order system, we have

Corollary 5.1.2   *Let $n_k \geq 0$ $(0 \leq k \leq n)$ be given integers and assume that a_k, $p_{k,j}$ $(0 \leq k \leq n, 0 \leq j \leq n_k)$ and $\tau_{k,j} \geq 0$ $(1 \leq k \leq n, 0 \leq j \leq n_k)$ are given*

real numbers. Then the equation

$$\sum_{k=1}^{n} \frac{d^k}{dt^k}\left[a_k x(t) - \sum_{j=1}^{n_k} p_{k,j} x(t - \tau_{k,j}) \right] = \sum_{j=1}^{n_0} p_{0,j} x(t - \tau_{0,j})$$

has a nonoscillating solution if and only if its characteristic equation

$$\sum_{k=1}^{n} \lambda^k \left[a_k - \sum_{j=1}^{n_k} p_{k,j}\, e^{-\lambda \tau_{k,j}} \right] = \sum_{j=1}^{n_0} p_{0,j}\, e^{-\lambda \tau_{0,j}}$$

has a real root.

5.2   Notes and Comments to Chapter 5

The results of Section 5.1 are due to Arino and Györi [5]. Sufficient conditions for oscillation of the solutions of linear systems with constant coefficients of neutral type were obtained by Györi and Ladas [65]. We shall note that the first work devoted to the investigation of oscillatory properties of the solutions of systems of differential equations of neutral type was by Shevelo et al [143].

6

Neutral partial differential equations

6.1 Linear parabolic differential equations

In the last few years the fundamental theory of partial differential equations with a translated argument has been developed intensively. The qualitative theory of these important classes of partial differential equations, however, is still in an initial stage of development. Thus, for instance, a small number of papers are devoted to the oscillation theory for this class of equations, published in the period after 1984. Sufficient conditions for oscillation of the solutions of linear parabolic differential equations with delay were obtained in the papers of Kreith and Ladas [92] and of neutral type—in the paper of Mishev and Bainov [120].

In this section sufficient conditions for the oscillation of the solutions of linear parabolic equations of neutral type of the form

$$\frac{\partial}{\partial t}\left[u - \sum_{i=1}^{m} \lambda_i(t)u(x, t - \tau_i) \right]$$

$$= a(t)\,\Delta u + \sum_{i=1}^{s} a_i(t)\,\Delta u(x, t - \rho_i) - p(x, t)u - \sum_{i=1}^{k} p_i(x, t)u(x, t - \sigma_i),$$

$$(x, t) \in \Omega \times (0, \infty) \equiv G \tag{6.1.1}$$

are obtained, where Ω is a bounded domain in $\mathbb{R}^n$ with a piecewise smooth boundary, $u = u(x, t)$, $\Delta u = \sum_{i=1}^{n} u_{x_i x_i}(x, t)$, τ_i, ρ_i, $\sigma_i = \text{const.} > 0$.

Consider boundary conditions of the form

$$\frac{\partial u}{\partial n} + \gamma(x, t)u = 0, \quad (x, t) \in \partial\Omega \times [0, \infty), \tag{6.1.2}$$

$$u = 0, \quad (x, t) \in \partial\Omega \times [0, \infty). \tag{6.1.3}$$

We shall say that conditions (H6.1) are satisfied if the following conditions hold:

H6.1.1 $\lambda_i(t) \in C^1([0, \infty); [0, \infty)), i = 1, 2, \ldots, m,$

$$\sum_{i=1}^{m} \lambda_i(t) \leq 1, \quad t \geq 0.$$

H6.1.2 $a(t), \quad a_i(t) \in C([0, \infty); [0, \infty)), i = 1, 2, \ldots, s.$
H6.1.3 $p(x, t) \in C(\bar{G}; [0, \infty)),$

$$p_i(x, t) \in C(\bar{G}; (0, \infty)), \quad i = 1, 2, \ldots, k.$$

H6.1.4 $\gamma(x, t) \in C(\partial\Omega \times [0, \infty); [0, \infty)).$

Definition 6.1.1 The solution $u(x, t) \in C^2(G) \cap C^1(\bar{G})$ of problem (6.1.1), (6.1.2), ((6.1.1), (6.1.3)) is said to oscillate in the domain $G = \Omega \times (0, \infty)$ if for any positive number μ there exists a point $(x_0, t_0) \in \Omega \times [\mu, \infty)$ such that the equality $u(x_0, t_0) = 0$ holds.

In the following theorems sufficient conditions are obtained for oscillation of the solutions of problems (6.1.1), (6.1.2) and (6.1.1), (6.1.3) in the domain G. In relation to this a class of neutral differential inequalities is investigated. We shall note that conditions for oscillation of the solutions of equation (6.1.1) for $\lambda_i(t) \equiv 0, \quad i = 1, 2, \ldots, m, a_i(t) \equiv 0, \quad i = 1, 2, \ldots, s, \quad p(x, t) \equiv 0$ and $k = 1$ were obtained in the papers of Kreith and Ladas [92] and Yoshida [156].

Introduce the following notation:

$$P(t) = \min\{p(x, t): x \in \bar{\Omega}\},$$
$$\tag{6.1.4}$$
$$P_i(t) = \min\{p_i(x, t): x \in \bar{\Omega}\}, \quad i = 1, 2, \ldots, k.$$

With each solution $u(x, t) \in C^2(G) \cap C^1(\bar{G})$ of problem (6.1.1), (6.1.2) we associate the function

$$v(t) = \int_{\Omega} u(x, t) \, dx, \quad t \geq 0. \tag{6.1.5}$$

Lemma 6.1.1

Let conditions (H6.1) hold and let $u(x, t)$ be a positive solution of problem (6.1.1), (6.1.2) in the domain G. Then the function $v(t)$ defined by (6.1.5) satisfies the differential inequality

$$\frac{d}{dt}\left[v(t) - \sum_{i=1}^{m} \lambda_i(t)v(t - \tau_i) \right] + P(t)v(t) + \sum_{i=1}^{k} P_i(t)v(t - \sigma_i) \leq 0, \quad t \geq t_0,$$

$$(6.1.6)$$

where t_0 is a sufficiently large positive number.

Proof

Let $u(x, t)$ be a positive solution of problem (6.1.1), (6.1.2) in the domain G and $t_0 = \max\{\tau_1, \ldots, \tau_m, \rho_1, \ldots, \rho_s, \sigma_1, \ldots, \sigma_k\}$. Then $u(x, t - \tau_i) > 0$, $u(x, t - \rho_i) > 0$ and $u(x, t - \sigma_i) > 0$ for $(x, t) \in \Omega \times [t_0, \infty)$. Integrate both sides of equation (6.1.1) with respect to x over the domain Ω and for $t \geq t_0$ obtain

$$\frac{d}{dt}\left[\int_{\Omega} u(x, t) \, dx - \sum_{i=1}^{m} \lambda_i(t) \int_{\Omega} u(x, t - \tau_i) \, dx \right]$$

$$= a(t) \int_{\Omega} \Delta u(x, t) \, dx + \sum_{i=1}^{s} a_i(t) \int_{\Omega} \Delta u(x, t - \rho_i) \, dx$$

$$- \int_{\Omega} p(x, t)u(x, t) \, dx - \sum_{i=1}^{k} \int_{\Omega} p_i(x, t)u(x, t - \sigma_i) \, dx. \quad (6.1.7)$$

From Green's formula and conditions H6.1.4 it follows that

$$\int_{\Omega} \Delta u(x, t) \, dx = \int_{\partial\Omega} \frac{\partial u}{\partial n} \, ds = - \int_{\partial\Omega} \gamma(x, t)u(x, t) \, ds \leq 0, \quad (6.1.8)$$

$$\int_{\Omega} \Delta u(x, t - \rho_i) \, dx = \int_{\partial\Omega} \frac{\partial u}{\partial n}(x, t - \rho_i) \, ds = - \int_{\partial\Omega} \gamma(x, t - \rho_i)u(x, t - \rho_i) \leq 0.$$

$$(6.1.9)$$

Moreover, from (6.1.4) it follows that

$$\int_{\Omega} p(x,t)u(x,t)\,dx \geq P(t)\int_{\Omega} u(x,t)\,dx = P(t)v(t), \qquad (6.1.10)$$

$$\int_{\Omega} p_i(x,t)u(x,t-\sigma_i)\,dx \geq P_i(t)\int_{\Omega} u(x,t-\sigma_i)\,dx = P_i(t)v(t-\sigma_i).$$

$$(6.1.11)$$

Using (6.1.8)–(6.1.11) and conditions H6.1.2, from (6.1.7) we obtain

$$\frac{d}{dt}\left[v(t) - \sum_{i=1}^{m} \lambda_i(t)v(t-\tau_i)\right] \leq -P(t)v(t) - \sum_{i=1}^{k} P_i(t)v(t-\sigma_i)$$

which proves Lemma 6.1.1.

Definition 6.1.2 The solution $v(t) \in C^1([t_0, \infty); \mathbb{R})$ of the differential inequality (6.1.6) is said to be eventually positive (negative) if there exists a sufficiently large positive number t_1 such that the inequality $v(t) > 0$ $(v(t) < 0)$ holds for $t \geq t_1$.

Theorem 6.1.1

Let conditions (H6.1) hold and let the differential inequality (6.1.6) have no eventually positive solutions. Then each solution $u(x,t)$ of problem (6.1.1), (6.1.2) oscillates in the domain G.

Proof

Let $\mu > 0$ be an arbitrary number. Suppose that the assertion is not true, that is, that $u(x,t)$ is a solution of problem (6.1.1), (6.1.2) having no zero in the domain $G_\mu = \Omega \times [\mu, \infty)$. If $u(x,t) > 0$ for $(x,t) \in G_\mu$, then from Lemma 6.1.1 it follows that the function $v(t)$ defined by (6.1.5) is a positive solution of inequality (6.1.6) for $t \geq t_0 + \mu$ which contradicts the condition of the theorem. If $u(x,t) < 0$ for $(x,t) \in G_\mu$, then $-u(x,t)$ is a positive solution of problem (6.1.1), (6.1.2). From Lemma 6.1.1 it follows that the function $-v(t) = -\int_{\Omega} u(x,t)\,dx$ is a positive solution of inequality (6.1.6) for $t \geq t_0 + \mu$ which also contradicts the condition of the theorem.

Now we shall investigate the oscillatory properties of the solutions of problem (6.1.1), (6.1.3). Consider in the domain the Dirichlet problem

$$\Delta U(x) + \alpha U(x) = 0, \quad x \in \Omega, \tag{6.1.12}$$

$$U(x) = 0, \quad x \in \partial\Omega \tag{6.1.13}$$

where $\alpha = \text{const}$. It is well known that the smallest eigenvalue α_0 of problem (6.1.12), (6.1.13) is positive and the corresponding eigenfunction $\varphi(x)$ can be chosen so that $\varphi(x) > 0$ for $x \in \Omega$, [150].

With each solution $u(x, t) \in C^2(G) \cap C^1(\bar{G})$ of problem (6.1.1), (6.1.3) we associate the function

$$w(t) = \int_\Omega u(x, t)\varphi(x)\,dx, \quad t \geq 0. \tag{6.1.14}$$

We shall note that such averaging was first used by Yoshida [156].

Lemma 6.1.2

Let conditions H6.1.1–H6.1.3 and let $u(x, t)$ be a positive solution of problem (6.1.1), (6.1.3) in the domain G. Then the function $w(t)$ defined by (6.1.14) satisfies the neutral differential inequality

$$\frac{d}{dt}\left[w(t) - \sum_{i=1}^m \lambda_i(t)w(t - \tau_i) \right] + [\alpha_0 a(t) + P(t)]w(t)$$

$$+ \alpha_0 \sum_{i=1}^s a_i(t)w(t - \rho_i) + \sum_{i=1}^k P_i(t)w(t - \sigma_i) \leq 0, \quad t \geq t_0 \tag{6.1.15}$$

where t_0 is a sufficiently large positive number.

Proof

Let $u(x, t)$ be a positive solution of problem (6.1.1), (6.1.3) in the domain G and $t_0 = \max\{\tau_1, \ldots, \tau_m, \rho_1, \ldots, \rho_s, \sigma_1, \ldots, \sigma_k\}$. Then $u(x, t - \tau_i) > 0$, $u(x, t - \rho_i) > 0$, $u(x, t - \sigma_i) > 0$ for $(x, t) \in \Omega \times [t_0, \infty)$. Multiply both sides of equation (6.1.1) by the eigenfunction $\varphi(x)$ and integrate with respect

x over the domain Ω. For $t \geqq t_0$ we obtain

$$\frac{d}{dt}\left[\int_\Omega u(x,t)\varphi(x)\,dx - \sum_{i=1}^m \lambda_i(t)\int_\Omega u(x,t-\tau_i)\varphi(x)\,dx\right]$$

$$= a(t)\int_\Omega \Delta u(x,t)\varphi(x)\,dx + \sum_{i=1}^s a_i(t)\int_\Omega \Delta u(x,t-\rho_i)\varphi(x)\,dx$$

$$- \int_\Omega p(x,t)u(x,t)\varphi(x)\,dx$$

$$- \sum_{i=1}^k \int_\Omega p_i(x,t)u(x,t-\sigma_i)\varphi(x)\,dx. \tag{6.1.16}$$

From Green's formula it follows that

$$\int_\Omega \Delta u(x,t)\varphi(x)\,dx = \int_\Omega u(x,t)\Delta\varphi(x)\,dx$$

$$= -\alpha_0 \int_\Omega u(x,t)\varphi(x)\,dx = -\alpha_0 w(t), \quad (6.1.17)$$

$$\int_\Omega \Delta u(x,t-\rho_i)\varphi(x)\,dx = \int_\Omega u(x,t-\rho_i)\Delta\varphi(x)\,dx$$

$$= -\alpha_0 \int_\Omega u(x,t-\rho_i)\varphi(x)\,dx = -\alpha_0 w(t-\rho_i)$$

$$\tag{6.1.18}$$

where $\alpha_0 > 0$ is the smallest eigenvalue of problem (6.1.12), (6.1.13). Moreover, from (6.1.4) it follows that

$$\int_\Omega p(x,t)u(x,t)\varphi(x)\,dx \geqq P(t)\int_\Omega u(x,t)\varphi(x)\,dx = P(t)w(t) \quad (6.1.19)$$

$$\int_{\Omega} p_i(x, t)u(x, t - \sigma_i)\varphi(x)\,dx$$

$$\geq P_i(t)\int_{\Omega} u(x, t - \sigma_i)\varphi(x)\,dx = -P_i(t)w(t - \sigma_i). \qquad (6.1.20)$$

Using (6.1.17)–(6.1.20) and conditions H6.1.2, from (6.1.16) we get

$$\frac{d}{dt}\left[w(t) - \sum_{i=1}^{m} \lambda_i(t)w(t - \tau_i)\right]$$

$$\leq -\alpha_0 a(t)w(t) - \alpha_0 \sum_{i=1}^{s} a_i(t)w(t - \rho_i) - P(t)w(t) - \sum_{i=1}^{k} P_i(t)w(t - \sigma_i).$$

This completes the proof of Lemma 6.1.2.
Analogously to Theorem 6.1.1 the following theorem is proved.

Theorem 6.1.2

Let conditions H6.1.1–H6.1.3 hold and let the differential inequality (6.1.15) have no eventually positive solutions. Then each solution $u(x, t)$ of problem (6.1.1), (6.1.3) oscillates in the domain G.

From the theorems proved above in this section it follows that the finding of sufficient conditions for oscillation of the solutions of equation (6.1.1) in the domain G is reduced to the investigation of the oscillatory properties of neutral differential inequalities of the form

$$\frac{d}{dt}\left[x(t) - \sum_{i=1}^{m} \lambda_i(t)x(t - \tau_i)\right] + q(t)x(t) + \sum_{i=1}^{k} q_i(t)x(t - \sigma_i) \leq 0, \quad t \geq t_0$$

$$(6.1.21)$$

$$\frac{d}{dt}\left[x(t) - \sum_{i=1}^{m} \lambda_i(t)x(t - \tau_i)\right] + q(t)x(t) + \sum_{i=1}^{k} q_i(t)x(t - \sigma_i) \geq 0, \quad t \geq t_0.$$

$$(6.1.22)$$

Together with (6.1.21) and (6.1.22) we shall consider the neutral differential equation

$$\frac{d}{dt}\left[x(t) - \sum_{i=1}^{m} \lambda_i(t)x(t - \tau_i)\right] + q(t)x(t) + \sum_{i=1}^{k} q_i(t)x(t - \sigma_i) = 0, \quad t \geq t_0.$$

$$(6.1.23)$$

Assume that the following conditions hold:

H6.1.5 $\lambda_i(t) \in C^1([t_0, \infty); [0, \infty))$, $i = 1, 2, \ldots, m$,

$$\sum_{i=1}^m \lambda_i(t) \le 1, \quad t \ge t_0 \ge 0.$$

H6.1.6 $q(t) \in C([t_0, \infty); [0, \infty))$.

H6.1.7 $q_i(t) \in C([t_0, \infty); (0, \infty))$, $i = 1, 2, \ldots, k$.

In the proof of the following theorem we shall use the following result of Ladas and Stavroulakis [103] concerning the differential inequality (6.1.21) in the case when $\lambda_i(t) \equiv 0$, $i = 1, 2, \ldots, m$ and $q(t) \equiv 0$.

Theorem 6.1.3 [103]

Let condition H6.1.7 hold in addition to the following conditions:

$$\liminf_{t \to \infty} \int_{t - \sigma_i/2}^{t} q_i(s)\,ds > 0, \quad i = 1, 2, \ldots, k,$$

$$\liminf_{t \to \infty} \int_{t - \sigma_i}^{t} q_i(s)\,ds > 1/e \quad \text{for at least one } i \in \{1, 2, \ldots, k\}.$$

Then the differential inequality

$$x'(t) + \sum_{i=1}^k q_i(t) x(t - \sigma_i) \le 0, \quad t \ge t_0$$

has no eventually positive solutions.

Theorem 6.1.4

Let conditions H6.1.5–H6.1.7 hold in addition to the following conditions:

$$q_i(t) \ge C = \text{const.} > 0, \quad t \ge t_0 \quad \text{for at least one } i \in \{1, 2, \ldots, k\}, \quad (6.1.24)$$

$$\liminf_{t \to \infty} \int_{t - \sigma_i/2}^{t} q_i(s)\,ds > 0, \quad i = 1, 2, \ldots, k, \quad (6.1.25)$$

$$\liminf_{t \to \infty} \int_{t-\sigma_i}^{t} q_i(s)\,ds > \frac{1}{e}\exp\left[-\liminf_{t \to \infty} \int_{t-\sigma_i}^{t} q(s)\,ds\right] \qquad (6.1.26)$$

for at least one $i \in \{1, 2, \ldots, k\}$.
   Then:
   (*i*)   *the differential inequality* (6.1.21) *has no eventually positive solutions;*
   (*ii*) *the differential inequality* (6.1.22) *has no eventually negative solutions;*
   (*iii*) *all solutions of the differential equation* (6.1.23) *oscillate.*

Proof

(*i*) Suppose that there exists an eventually positive solution $x(t)$ of the differential inequality (6.1.21). Hence $x(t) > 0$ for $t \geq t_1$, where $t_1 \geq t_0$.
Introduce the notation

$$z(t) = x(t) - \sum_{i=1}^{m} \lambda_i(t)x(t - \tau_i), \quad t \geq t_1 + \tau,$$

$$(6.1.27)$$

$$\tau = \max\{\tau_1, \ldots, \tau_m\}, \quad \sigma = \max\{\sigma_1, \ldots, \sigma_k\}.$$

From conditions H6.1.6, H6,1,7 and (6.1.21) it follows that $z'(t) < 0$ for $t \geq t_2$, where $t_2 = t_1 + \tau + \sigma$, that is, the function $z(t)$ is strictly decreasing in the interval $[t_2, \infty)$. We shall prove that the function $z(t)$ is bounded below. Suppose that $\lim_{t \to \infty} z(t) = -\infty$. Then from (6.1.27), using conditions H6.1.5, we obtain that the function $x(t)$ is not bounded. Hence there exists a number $t_3 \geq t_2 + \tau$ such that

$$z(t_3) < 0 \quad \text{and} \quad x(t_3) = \max_{s \in [t_2, t_3]} x(s) \qquad (6.1.28)$$

We use condition H6.1.5 and obtain

$$z(t_3) = x(t_3) - \sum_{i=1}^{m} \lambda_i(t_3)x(t_3 - \tau_i) \geq x(t_3)\left[1 - \sum_{i=1}^{m} \lambda_i(t_3)\right] \geq 0.$$

The inequality obtained contradicts the inequality in (6.1.28). Hence the function $z(t)$ is bounded below in the interval $[t_3, \infty)$ and let L be its greatest lower bound. Since $z(t)$ is a decreasing function, then

$$\lim_{t \to \infty} z(t) = L \tag{6.1.29}$$

Integrate the differential inequality $(6.1.21)$ over the interval $[t_3, t]$, $t > t_3$ and using condition $(6.1.24)$ obtain

$$0 < C \int_{t_3}^{t} x(s - \sigma_{i_0}) \, ds \leq \int_{t_3}^{t} \left[q(s)x(s) + \sum_{i=1}^{k} q_i(s)x(s - \sigma_i) \right] ds$$

$$\leq z(t_3) - z(t).$$

Hence $x(t) \in L_1(t_0, \infty)$. From conditions H6.1.5 it follows that $z(t) \in L_1(t_0, \infty)$ and from $(6.1.29)$ it follows that $\lim_{t \to \infty} z(t) = 0$. Since $z(t)$ is a strictly decreasing function, then

$$z(t) > 0 \quad \text{for} \quad t \geq t_4 \tag{6.1.30}$$

where $t_4 \geq t_3$ is a sufficiently large number. Using the inequality $z(t) \leq x(t)$, $t \geq t_4$, from $(6.1.21)$ we obtain

$$z'(t) + q(t)z(t) + \sum_{i=1}^{k} q_i(t)z(t - \sigma_i) \leq 0, \quad t \geq t_4. \tag{6.1.31}$$

In the differential inequality $(6.1.31)$ we perform a change of the unknown function

$$y(t) = z(t) \exp\left[\int_{t_4}^{t} q(s) \, ds \right].$$

From $(6.1.30)$ it follows that

$$y(t) > 0 \quad \text{for} \quad t \geq t_4. \tag{6.1.32}$$

Moreover, $y(t)$ satisfies the differential inequality

$$y'(t) + \sum_{i=1}^{k} Q_i(t)y(t - \sigma_i) \leq 0, \quad t \geq t_4 \tag{6.1.33}$$

where $Q_i(t) = q_i(t) \exp[\int_{t-\sigma_i}^{t} q(s) \, ds]$. In view of conditions $(6.1.25)$ and $(6.1.26)$, for the coefficients $Q_i(t)$ of the differential inequality $(6.1.33)$ we obtain

$$\liminf_{t\to\infty} \int_{t-\sigma_i/2}^{t} Q_i(s)\,ds \geq \liminf_{t\to\infty} \int_{t-\sigma_i/2}^{t} q_i(s)\,ds\, \exp\left[\liminf_{t\to\infty} \int_{t-\sigma_i}^{t} q(s)\,ds\right] > 0,$$

$$i = 1, 2, \ldots, k.$$

$$\liminf_{t\to\infty} \int_{t-\sigma_i}^{t} Q_i(s)\,ds \geq \liminf_{t\to\infty} \int_{t-\sigma_i}^{t} q_i(s)\,ds\, \exp\left[\liminf_{t\to\infty} \int_{t-\sigma_i}^{t} q(s)\,ds\right]$$

$$> \frac{1}{e}\exp\left[-\liminf_{t\to\infty} \int_{t-\sigma_i}^{t} q(s)\,ds\right]\exp\left[\liminf_{t\to\infty} \int_{t-\sigma_i}^{t} q(s)\,ds\right]$$

$$= 1/e$$

for at least one $i \in \{1, 2, \ldots, k\}$.

Hence, for the differential inequality (6.1.33) all conditions of Theorem 6.1.3 are met, from which we obtain that (6.1.33) has no eventually positive solutions. But from (6.1.32) it follows that $y(t)$ is an eventually positive solution of (6.1.33).

Thus assertion (i) of Theorem 3.1.4 is proved.

(ii) The proof follows from the fact that if $x(t)$ is an eventually negative solution of the differential inequality (6.1.22), then $-x(t)$ is an eventually positive solution of the differential inequality (6.1.21).

(iii) From (i) and (ii) it follows that (6.1.23) has no eventually positive and eventually negative solutions. Hence all solutions of the differential equation (6.1.23) oscillate.

Remark 6.1.1 A question arises whether condition (6.1.24) of Theorem 6.1.4 is a consequence of condition (6.1.25) of the same theorem. The following example answers this question negatively. Consider the function $f(t) = \cos^2 t + e^{-t}\sin^2 t, \quad t \geq t_0$. A straightforward verification yields that

$$\liminf_{t\to\infty} \int_{t-2\pi}^{t} f(s)\,ds \geq \liminf_{t\to\infty} \int_{t-2\pi}^{t} \cos^2 s\,ds = \pi > 0.$$

On the other hand,

$$\lim_{k \to \infty} f\left(\frac{\pi}{2} + k\pi\right) = \lim_{k \to \infty} e^{-\frac{\pi}{2}(2k+1)} = 0$$

which implies that condition (6.1.24) is not satisfied for the function $f(t)$.

Remark 6.1.2 In the proof of Theorem 6.1.4 the following auxiliary assertion was used: if $f(t), g(t) \in C([t_0, \infty); \mathbb{R})$ and $f(t) \geq 0, t \geq t_0$, then

$$\liminf_{t \to \infty} \int_{t-\tau}^{t} f(s)g(s)\, ds \geq \liminf_{t \to \infty} g(t) \liminf_{t \to \infty} \int_{t-\tau}^{t} f(s)\, ds.$$

In fact, from the mean value theorem it follows that

$$\int_{t-\tau}^{t} f(s)g(s)\, ds = g(\eta_t) \int_{t-\tau}^{t} f(s)\, ds, \quad t - \tau \leq \eta_t \leq t.$$

Then

$$\liminf_{t \to \infty} \int_{t-\tau}^{t} f(s)g(s)\, ds = \liminf_{t \to \infty} g(\eta_t) \liminf_{t \to \infty} \int_{t-\tau}^{t} f(s)\, ds$$

$$\geq \liminf_{t \to \infty} g(t) \liminf_{t \to \infty} \int_{t-\tau}^{t} f(s)\, ds$$

since the special sequence $\eta_t \to \infty$ as $t \to \infty$.

A corollary of Theorem 6.1.1 and 6.1.4 is the following sufficient condition for oscillation of the solutions of problem (6.1.1), (6.1.2).

Theorem 6.1.5

Let conditions H6.1 hold in addition to the following conditions:

$$P_i(t) \geq C = \text{const.} > 0 \quad \text{for at least one } i \in \{1, 2, \ldots, k\} \quad (6.1.34)$$

$$\liminf_{t \to \infty} \int_{t-\sigma_i/2}^{t} P_i(s)\, ds > 0, \quad i = 1, 2, \ldots, k \tag{6.1.35}$$

$$\liminf_{t \to \infty} \int_{t-\sigma_i}^{t} P_i(s)\, ds > \frac{1}{e}\left[-\liminf_{t \to \infty} \int_{t-\sigma_i}^{t} P(s)\, ds\right] \tag{6.1.36}$$

for at least one $i \in \{1, 2, \ldots, k\}$.
Then each solution $u(x, t)$ of problem (6.1.1), (6.1.2) oscillates in G.

A corollary of Theorem 6.1.2 and Theorem 6.1.4 is the following sufficient condition for oscillation of the solutions of problem (6.1.1), (6.1.3).

Theorem 6.1.6

Let conditions H6.1.1–H6.1.3 hold as well as the following conditions:

$$a_i(t) > 0, \quad t \geq t_0, \quad i = 1, 2, \ldots, s \tag{6.1.37}$$

$$a_i(t) \geq C = \text{const.} > 0 \quad \text{for at least one } i \in \{1, 2, \ldots, s\} \tag{6.1.38}$$

or

$$P_i(t) \geq C = \text{const.} > 0 \quad \text{for at least one } i \in \{1, 2, \ldots, k\}.$$

$$\liminf_{t \to \infty} \int_{t-\sigma_i/2}^{t} P_i(s)\, ds > 0, \quad i = 1, 2, \ldots, k. \tag{6.1.39}$$

$$\liminf_{t \to \infty} \int_{t-\rho_i/2}^{t} a_i(s)\, ds > 0, \quad i = 1, 2, \ldots, s \tag{6.1.40}$$

$$\liminf_{t \to \infty} \int_{t-\rho_i}^{t} a_i(s)\, ds > A/\alpha_0 \quad \text{for at least one } i \in \{1, 2, \ldots, s\} \tag{6.1.41}$$

or

$$\liminf_{t \to \infty} \int_{t-\sigma_i}^{t} P_i(s)\, ds > B \quad \text{for at least one } i \in \{1, 2, \ldots, k\}$$

where

$$A = \frac{1}{e}\exp\left[-\lim_{t\to\infty}\inf \int_{t-\rho_i}^{t} (\alpha_0 a(s) + P(s))\,ds\right],$$

$$B = \frac{1}{e}\exp\left[-\lim_{t\to\infty}\inf \int_{t-\sigma_i}^{t} (\alpha_0 a(s) + P(s))\,ds\right].$$

Then each solution $u(x, t)$ of problem (6.1.1), (6.1.3) oscillates in G.

Example 6.1.1 Consider the equation

$$\frac{\partial}{\partial t}[u - u(x, t - \pi)] = u_{xx} + 2u_{xx}(x, t - \pi/2) - u(x, t) - 2u(x, t - \pi),$$

$$(6.1.42)$$

$$(x, t) \in (0, \pi) \times (0, \infty) \equiv G$$

with boundary condition

$$u_x(0, t) = u_x(\pi, t) = 0, \quad t \geq 0. \tag{6.1.43}$$

A straightforward verification shows that the functions $\lambda_1(t) \equiv 1$, $a(t) \equiv 1$, $a_1(t) \equiv 2$, $p(x, t) \equiv 1$, $p_1(x, t) \equiv 2$, $\gamma(x, t) \equiv 0$ satisfy all conditions of Theorem 6.1.5. Hence all the solutions of problem (6.1.42), (6.1.43) oscillate in the domain G. For instance, the function $u(x, t) = \sin t \cos x$ is such a solution.

Example 6.1.2 Consider the equation

$$\frac{\partial}{\partial t}[u(x, t) - e^{-2\pi} u(x, t - 2\pi)] = u_{xx} + e^{-\pi}u_{xx}(x, t - \pi)$$

$$- u - e^{-\pi} u(x, t - \pi), \quad (x, t) \in (0, \pi) \times (0, \infty) \equiv G$$

$$(6.1.44)$$

with boundary condition

$$u(0, t) = u(\pi, t) = 0, \quad t \geq 0. \tag{6.1.45}$$

A straightforward verification shows that the functions $\lambda_1(t) \equiv e^{-2\pi}$, $a(t) \equiv 1$, $a_1(t) = e^{-\pi}$, $p(x, t) \equiv 1$, $p_1(x, t) \equiv e^{-\pi}$, $\alpha_0 = 1$ satisfy all conditions of Theorem 6.1.6. Hence all solutions of problem (6.1.44), (6.1.45) oscillate in the domain G. For instance, the function $u(x,t) = e^{-t} \cos t \sin x$ is such a solution.

6.2   Nonlinear parabolic differential equations

Sufficient conditions for oscillation of the solutions of parabolic differential equations with a deviating argument were obtained in the works of Yoshida [153], [156], Bykov and Kultaev [17] and Mishev [115]. We shall specially note the work of Yoshida [156] in which sufficient conditions for oscillation of the solutions of equations of the form

$$u_t = a(t)\,\Delta u \pm q(x, t)f(u(x, \sigma(t))), \quad (x, t) \in \Omega \times [0, \infty)$$

are found, where Ω is a bounded domain in $\mathbb{R}^n$.

In this section sufficient conditions for oscillation of the solutions of neutral nonlinear parabolic equations of the form

$$\frac{\partial}{\partial t}[u(x, t) + \lambda(t)u(x, t - \tau)] = a(t)\,\Delta u(x, t) - p(x, t)f(u(x, t - \tau)),$$

$$(6.2.1)$$

$$(x, t) \in \Omega \times (t_0, \infty) \equiv G, \quad t_0 \geq 0$$

are found, where $\tau = \text{const.} > 0$, $\Delta u(x, t) = \sum_{i=1}^{n} u_{x_i x_i}(x, t)$ and Ω is a bounded domain in $\mathbb{R}^n$ with a piecewise smooth boundary. Consider two types of boundary conditions:

$$\frac{\partial u}{\partial n} + \gamma(x, t)u = 0, \quad (x, t) \in \partial\Omega \times [t_0, \infty) \qquad (6.2.2)$$

$$u = 0, \quad (x, t) \in \partial\Omega \times [t_0, \infty). \qquad (6.2.3)$$

Definition 6.2.1   The solution $u(x, t) \in C^2(G) \cap C^1(\bar{G})$ of problem (6.2.1), (6.2.2) ((6.2.1), (6.2.3)) is said to oscillate in the domain G if for any number $\mu \geq t_0$ there exists a point $(x_1, t_1) \in G_\mu = \Omega \times [\mu, \infty)$ such that the condition $u(x_1, t_1) = 0$ holds.

We shall say that conditions (H6.2) are met if the following conditions hold:

H6.2.1 $a(t) \in C([t_0, \infty); [0, \infty))$.

H6.2.2 $p(x, t) \in (\bar{G}; [0, \infty))$.

H6.2.3 $\lambda(t) \in C^1([t_0, \infty); \mathbb{R})$

$$0 < \lambda_1 \leq \lambda(t) \leq \lambda_2 \quad \text{for} \quad t \geq t_0, \quad \lambda_1, \lambda_2 = \text{const.}$$

H6.2.4 $f(u) \in C(\mathbb{R}; \mathbb{R})$, $f(-u) = -f(u)$.

$f(u)$ is a positive monotone increasing and convex function in the interval $(0, \infty)$.

H6.2.5 $\gamma(x, t) \in C(\partial\Omega \times [t_0, \infty); [0, \infty))$.

We shall note that sufficient conditions for oscillation of the solutions of problems (6.2.1), (6.2.2) and (6.2.1), (6.2.3) in the case when $\lambda(t) \equiv 0$ were obtained by Yoshida [156].

Introduce the following notation:

$$P(t) = \min\{p(x, t) : x \in \bar{\Omega}\}. \tag{6.2.4}$$

With each solution $u(x, t)$ of problem (6.2.1), (6.2.2) we associate the function

$$v(t) = \int_\Omega u(x, t)\, dx \left(\int_\Omega dx \right)^{-1}, \quad t \geq t_0 \tag{6.2.5}$$

Lemma 6.2.1

Let conditions (H6.2) hold and let $u(x, t)$ be a positive solution in the domain G of problem (6.2.1), (6.2.2). Then the function $v(t)$ defined by (6.2.5) satisfies the neutral differential inequality

$$\frac{d}{dt}[v(t) + \lambda(t)v(t - \tau)] + P(t)f(v(t - \tau)) \leq 0,$$

$$\tag{6.2.6}$$

$$t \geq t_0 + \tau.$$

Proof

Let $u(x, t)$ be a positive solution in the domain $G = \Omega \times [t_0, \infty)$ of problem (6.2.1), (6.2.2). Then $u(x, t - \tau) > 0$ for $(x, t) \in \Omega \times [t_0 + \tau, \infty)$. Integrate both sides of equation (6.2.1) with respect to x over the domain Ω and for $t \geq t_0 + \tau$ obtain

$$\frac{d}{dt}\left[\int_\Omega u(x,t)\,dx + \lambda(t)\int_\Omega u(x,t-\tau)\,dx\right]$$

$$= a(t)\int_\Omega \Delta u(x,t)\,dx - \int_\Omega p(x,t)f(u(x,t-\tau))\,dx. \qquad (6.2.7)$$

From Green's formula and condition H6.2.5 it follows that

$$\int_\Omega \Delta u(x,t)\,dx = \int_{\partial\Omega}\frac{\partial u}{\partial n}(x,t)\,ds = -\int_{\partial\Omega}\gamma(x,t)u(x,t)\,ds \leq 0. \qquad (6.2.8)$$

Moreover, from (6.2.4) and Jensen's inequality it follows that

$$\int_\Omega p(x,t)f(u(x,t-\tau))\,dx \geq P(t)\int_\Omega f(u(x,t-\tau))\,dx$$

$$\geq P(t)f\left(\int_\Omega u(x,t-\tau)\,dx\left(\int_\Omega dx\right)^{-1}\right)\int_\Omega dx$$

$$= P(t)f(v(t-\tau))\int_\Omega dx \qquad (6.2.9)$$

Using (6.2.8), (6.2.9) and condition H6.2.2, from (6.2.7) we obtain

$$\frac{d}{dt}[v(t)+\lambda(t)v(t-\tau)] \leq -P(t)f(v(t-\tau))$$

which proves Lemma 6.2.1.

Definition 6.2.2 The solution $v(t)\in C^1([t_0,\infty);\mathbb{R})$ of the differential inequality (6.2.6) is said to be eventually positive (negative) if there exists a sufficiently large positive number t_1 such that the inequality $v(t)>0$ ($v(t)<0$) holds for all $t \geq t_1$.

Theorem 6.2.1

Let conditions (H6.2) hold and let each eventually positive solution of the differential inequality (6.2.6) tend to zero as $t \to \infty$. Then each solution $u(x, t)$ of problem (6.2.1), (6.2.2) either oscillates in the domain G or the following equality holds

$$\lim_{t \to \infty} \int_{\Omega} u(x, t) \, \mathrm{d}x = 0.$$

Proof

Suppose that this is not true, that is, that the function $u(x, t)$ is a solution of problem (6.2.1), (6.2.2) without a zero in the domain $G_\mu = \Omega \times [\mu, \infty)$, $\mu \geq t_0$ and $\int_{\Omega} u(x, t) \, \mathrm{d}x \nrightarrow 0, t \to \infty$. Without loss of generality we can assume that $u(x, t) > 0$ for $(x, t) \in G_\mu$. Then from Lemma 6.2.1 it follows that the function $v(t)$ defined by (6.2.5) is a positive solution of inequality (6.2.6) for $t \geq \mu + \tau$ such that $v(t) \nrightarrow 0$, $t \to \infty$ which contradicts the condition of the theorem.

Now we shall investigate the oscillatory properties of the solutions of problem (6.2.1), (6.2.3). Let α_0 be the least eigenvalue and let $\varphi(x)$ be the corresponding eigenfunction of problem (6.1.12), (6.1.13). It is well known [150] that $\alpha_0 > 0$ and $\varphi(x)$ can be chosen so that $\varphi(x) > 0$ for $x \in \Omega$.

With each solution $u(x, t) \in C^2(G) \cap C^1(\bar{G})$ of problem (6.2.1), (6.2.3) we associate the function

$$w(t) = \int_{\Omega} u(x, t)\varphi(x) \, \mathrm{d}x \left(\int_{\Omega} \varphi(x) \, \mathrm{d}x \right)^{-1}, \quad t \geq t_0. \qquad (6.2.10)$$

Lemma 6.2.2

Let conditions H6.2.1–H6.2.4 hold and let $u(x, t)$ be a positive solution in the domain G of problem (6.2.1), (6.2.3). Then the function $w(t)$ defined by (6.2.10) satisfies the neutral differential inequality

$$\frac{\mathrm{d}}{\mathrm{d}t}[w(t) + \lambda(t)w(t - \tau)] + \alpha_0 a(t)w(t) + P(t)f(w(t - \tau)) \leq 0,$$

$$(6.2.11)$$

$$t \geq t_0 + \tau.$$

Proof

Let $u(x, t)$ be a positive solution in the domain $G = \Omega \times [t_0, \infty)$ of problem (6.2.1), (6.2.3). Then $u(x, t - \tau) > 0$ for $(x, t) \in \Omega \times [t_0 + \tau, \infty)$. Multiply both sides of equation (6.2.1) by the eigenfunction $\varphi(x)$ and integrate with respect to x over the domain Ω. For $t \geq t_0 + \tau$ we obtain

$$\frac{d}{dt} \left[\int_\Omega u(x, t) \varphi(x) \, dx + \lambda(t) \int_\Omega u(x, t - \tau) \varphi(x) \, dx \right]$$

$$= a(t) \int_\Omega \Delta u(x, t) \varphi(x) \, dx - \int_\Omega p(x, t) f(u(x, t - \tau)) \varphi(x) \, dx. \quad (6.2.12)$$

From Green's formula it follows that

$$\int_\Omega \Delta u(x, t) \varphi(x) \, dx = \int_\Omega u(x, t) \Delta \varphi(x) \, dx = -\alpha_0 \int_\Omega u(x, t) \varphi(x) \, dx$$

$$= -\alpha_0 w(t) \int_\Omega \varphi(x) \, dx. \quad (6.2.13)$$

Moreover, from (6.2.4) and Jensen's inequality it follows that

$$\int_\Omega p(x, t) f(u(x, t - \tau)) \varphi(x) \, dx$$

$$\geq P(t) \int_\Omega f(u(x, t - \tau)) \varphi(x) \, dx$$

$$\geq P(t) f \left(\int_\Omega u(x, t - \tau) \varphi(x) \, dx \left(\int_\Omega \varphi(x) \, dx \right)^{-1} \right) \int_\Omega \varphi(x) \, dx$$

$$= P(t) f(w(t - \tau)) \int_\Omega \varphi(x) \, dx \quad (6.2.14)$$

Using (6.2.13), (6.2.14) and condition H6.2.2, from (6.2.12) we obtain

$$\frac{d}{dt}[w(t) + \lambda(t)w(t - \tau)] \leq -\alpha_0 a(t)w(t) - P(t)f(w(t - \tau)).$$

This completes the proof of Lemma 6.2.2.
Analogously to Theorem 6.2.1 the following theorem is proved.

Theorem 6.2.2

Let conditions H6.2.1–H6.2.4 hold and let each eventually positive solution of the differential inequality (6.2.12) tend to zero as $t \to \infty$. Then each solution $u(x, t)$ of problem (6.2.1), (6.2.3) either oscillates in the domain G or the equality $\lim_{t \to \infty} \int_\Omega u(x, t)\varphi(x)\,dx = 0$ holds.

From the theorems proved above in this section it follows that the finding of sufficient conditions for oscillation of the solutions of equation (6.2.1) in the domain G is reduced to the investigation of the oscillatory properties of neutral nonlinear differential inequalities of the form

$$\frac{d}{dt}[x(t) + \lambda(t)x(t - \tau)] + q_0(t)x(t) + q(t)f(x(t - \tau)) \leq 0, \quad t \geq t_0$$

$$(6.2.15)$$

$$\frac{d}{dt}[x(t) + \lambda(t)x(t - \tau)] + q_0(t)x(t) + q(t)f(x(t - \tau)) \geq 0, \quad t \geq t_0.$$

$$(6.2.16)$$

Together with (6.2.15) and (6.2.16) consider the neutral nonlinear differential equation

$$\frac{d}{dt}[x(t) + \lambda(t)x(t - \tau)] + q_0(t)x(t) + q(t)f(x(t - \tau)) = 0, \quad t \geq t_0.$$

$$(6.2.17)$$

Definition 6.2.3 The solution $x(t)$ of the differential equation (6.2.17) is said to oscillate if there exists a sequence of zeros $\{t_n\}_{n=1}^\infty$ of the function $x(t)$ for which the equality $\lim_{n \to \infty} t_n = +\infty$ holds.
Assume the following conditions fulfilled:
H6.2.6 $q_0(t)$, $q(t) \in C([t_0, \infty); [0, \infty))$.
H6.2.7 $f(u) \in C(\mathbb{R}; \mathbb{R})$, $f(-u) = -f(u)$, where $f(u)$ is a positive and monotone increasing function in the interval $(0, \infty)$.

In the proof of the subsequent theorem we shall use the following lemma which in the case $\lambda(t) = \lambda = \text{const.} > 0$ was proved by Zahariev and Bainov [160].

Lemma 6.2.3

Let the following conditions hold:
(1) The function $x(t) \in C([t_0, \infty); [0, \infty))$.
(2) The function $\lambda(t) \in C([t_0, \infty); \mathbb{R})$,

$$0 < \lambda_1 \leqq \lambda(t) \leqq \lambda_2 \quad for \quad t \geqq t_0, \quad \lambda_1, \lambda_2 = \text{const.}$$

(3) For $t \geqq t_0$ the inequality

$$x(t) + \lambda(t)x(t - \tau) \geqq C$$

holds, where $\tau > 0$, $C > 0$.

Then there exists a set $E \subset [t_0, \infty)$ and a constant $C_1 > 0$ such that $x(t) \geqq C_1$ for $t \in E$ and $\text{meas}(E \cap [t, t + 2\tau]) \geqq \tau$ for $t \geqq t_0$.

Proof

Let $s \in [t_0, \infty)$ and consider the sets

$$A = \{t : s + \tau \leqq t \leqq s + 2\tau, \quad x(t) \geqq C/2\}$$

$$B = \{t : s + \tau \leqq t \leqq s + 2\tau, \quad x(t - \tau) \geqq C/2\lambda_2\}.$$

From the conditions of the lemma it follows that $A \cup B = [s + \tau, s + 2\tau]$. Then

$$\text{meas}\{t : s + \tau \leqq t \leqq s + 2\tau, \quad x(t) \geqq C/2\}$$
$$+ \text{meas}\{t : s \leqq t \leqq s + \tau, \quad x(t) \geqq C/2\lambda_2\}$$
$$= \text{meas } A + \text{meas } B \geqq \text{meas}(A \cup B) = \tau.$$

Choose $C_1 = \min\{C/2, C/2\lambda_2\}$, with which Lemma 6.2.3 is proved.

Theorem 6.2.3

Let the following conditions be satisfied:
(1) Conditions H6.2.6–H6.2.7 hold.

(2) *The function* $\lambda(t) \in C^1([t_0, \infty); \mathbb{R})$,

$$0 < \lambda_1 \leq \lambda(t) \leq \lambda_2 \quad for \quad t \geq t_0, \quad \lambda_1, \lambda_2 = \text{const.}$$

(3) *For any closed and measurable set* $E \subset [t_0, \infty)$ *for which* $\text{meas}(E \cap [t, t + 2\tau]) \geq \tau$, $t \in [t_0, \infty)$ *the following condition holds*

$$\int_E q(t) \, dt = \infty. \tag{6.2.18}$$

Then:

(*i*) *each eventually positive solution of the differential inequality* (6.2.15) *tends to zero as* $t \to \infty$;

(*ii*) *each eventually negative solution of the differential inequality* (6.2.16) *tends to zero as* $t \to \infty$;

(*iii*) *each solution of the differential equation* (6.2.17) *either oscillates or tends to zero as* $t \to \infty$.

Proof

(*i*) Let $x(t)$ be an eventually positive solution of the differential inequality (6.2.15). Then there exists a number $t_1 \geq t_0$ such that $x(t) > 0$ and $x(t - \tau) > 0$ for $t \geq t_1$. From conditions H6.2.6–H6.2.7 and (6.2.15) it follows that $d/dt[x(t) + \lambda(t)x(t - \tau)] \leq 0$ for $t \geq t_1$. Hence the function $x(t) + \lambda(t)x(t - \tau)$ is monotone decreasing in the interval $[t_1, \infty)$. Since $x(t) + \lambda(t)x(t - \tau) > 0$ in the same interval, then

$$\lim_{t \to \infty} [x(t) + \lambda(t)x(t - \tau)] = C_1 \geq 0. \tag{6.2.19}$$

Suppose that $\lim_{t \to \infty}[x(t) + \lambda(t)x(t - \tau)] = C_1 > 0$. Then $x(t) + \lambda(t)x(t - \tau) \geq C_1/2 > 0$ for $t \geq t_2 \geq t_1$. From Lemma 6.2.3 it follows that there exists a closed and measurable set $E \subset [t_2, \infty)$ and a constant $C_2 > 0$ such that $x(t - \tau) \geq C_2$ for $t \in E$ and $\text{meas}(E \cap [t, t + 2\tau]) \geq \tau$ for $t \geq t_2$. Then from condition H6.2.7 it follows that $f(x(t - \tau)) \geq f(C_2) = C_3 > 0$ for $t \in E$. Integrate (6.2.15) over the interval $[t_2, t]$, $t_2 < t$ and obtain

$$C_3 \int_{E \cap [t_2, t]} q(s) \, ds \leq \int_{t_2}^{t} q(s) f(x(s - \tau)) \, ds$$

$$\leq x(t_2) + \lambda(t_2)x(t_2 - \tau) - [x(t) + \lambda(t)x(t - \tau)]$$

$$- \int_{t_2}^{t} q_0(s)x(s) \, ds \leq x(t_2) + \lambda(t_2)x(t_2 - \tau) = C_4.$$

For $t \to \infty$ from the above inequality it follows that $\int_E q(t)\,dt < \infty$ which contradicts condition (6.2.18). From (6.2.19) we obtain that $C_1 = 0$, that is, $\lim_{t \to \infty} [x(t) + \lambda(t)x(t - \tau)] = 0$ which implies that $\lim_{t \to \infty} x(t) = 0$.

Thus assertion (*i*) of Theorem 6.2.3 is proved.

(*ii*) The proof follows immediately from the fact that if $x(t)$ is an eventually negative solution of the differential inequality (6.2.16), then $-x(t)$ is an eventually positive solution of the differential inequality (6.2.15).

(*iii*) The proof follows immediately from assertions (*i*) and (*ii*).

A corollary of Theorems 6.2.1, 6.2.2 and 6.2.3 is the following sufficient condition for oscillation of the solutions of problems (6.2.1), (6.2.2) and (6.2.1), (6.2.3).

Theorem 6.2.4

Let the following conditions be fulfilled:

(1) Conditions (H6.2) hold.

(2) For any closed and measurable set $E \subset [t_0, \infty)$ for which $\mathrm{meas}(E \cap [t, t + 2\tau]) \geqq \tau, \quad t \in [t_0, \infty)$ *the following condition holds*

$$\int_E P(t)\,dt = \infty. \tag{6.2.20}$$

Then each solution $u(x, t)$ of problem (6.2.1), (6.2.2) ((6.2.1), (6.2.3)) either oscillates in the domain G or the following equality holds

$$\lim_{t \to \infty} \int_\Omega u(x, t)\,dx = 0 \quad \left(\lim_{t \to \infty} \int_\Omega u(x, t)\varphi(x)\,dx = 0 \right).$$

In the work of Yoshida [156] it was shown that for $\lambda(t) \equiv 0$ the result of Theorem 6.2.4 still holds if condition (6.2.20) is replaced by the weaker condition

$$\int_{t_0}^{\infty} P(t)\,dt = \infty. \tag{6.2.21}$$

It is of interest to establish whether condition (6.2.20) of Theorem 6.2.4 is essential or if it can in general be replaced by condition (6.2.21).

Example 6.2.1 Consider the following problem:

$$u_t(x, t) + u_t(x, t - \pi) = a(t) \Delta u - p(x, t)[u(x, t - \pi)]^3,$$
$$(6.2.22)$$
$$(x, t) \in (0, 1) \times (t_0, \infty), \quad t_0 > 0,$$
$$u_x(0, t) = u_x(1, t) = 0, \quad t \geq t_0 \qquad (6.2.23)$$

where $a(t) \in C([t_0, \infty); [0, \infty))$ is an arbitrary function and

$$p(x, t) = [t^{-2} + (t + \pi)^{-2}][t^{-1} + 1 - \cos t]^{-3}, \quad t \geq t_0.$$

In this case $f(u) = u^3$, $P(t) = \min\{p(x, t) : x \in [0, 1]\} = p(x, t)$. Introduce the notation $k_0 = [t_0] + 1$, $\Delta_k = [2k\pi - 1/k, 2k\pi + 1/k]$, where $k = 1, 2, 3, \ldots$. Straightforward calculations give us that

$$\int_{t_0}^{\infty} P(t) \, dt \geq \sum_{k=k_0}^{\infty} \int_{\Delta_k} (t^2 + (t + \pi)^2) t^{-2} (t + \pi)^{-2} (t^{-1} + 1 - \cos t)^{-3} \, dt$$

$$\geq \sum_{k=k_0}^{\infty} 4k^{-1} (2k\pi + k^{-1} + \pi)^{-2} \left(\frac{1}{2k\pi - 1/k} + 1 - \cos \frac{1}{k} \right)^{-3}$$

$$= \sum_{k=k_0}^{\infty} \frac{4k(2k^2\pi - 1)^2}{(2k^2\pi + k\pi + 1)^2 [k + (2k^2\pi - 1)\left(1 - \cos \frac{1}{k}\right)]^3}$$

which implies the relation $\int_{t_0}^{\infty} P(t) \, dt = \infty$.
On the other hand, if we set

$$E = \bigcup_{k=k_0}^{\infty} \left\{ t \geq t_0 : \frac{\pi}{4} + 2k\pi \leq t \leq \frac{5\pi}{4} + 2k\pi \right\}$$

then we obtain

$$\int_{E} P(t) \, dt \leq \int_{E} \frac{t^2 + (t + \pi)^2}{t^2 (t + \pi)^2 \left(t^{-1} + 1 - \frac{\sqrt{2}}{2} \right)^3} \, dt < \infty.$$

Hence the function $P(t)$ satisfies condition (6.2.21) but it does not satisfy condition (6.2.20) of Theorem 6.2.4. It is immediately verified that the function $u(x, t) = 1/(t + \pi) + 1 + \cos t$ is a solution of problem (6.2.22),

(6.2.23) which does not oscillate nor is the equality $\lim_{t \to \infty} \int_0^1 u(x, t) \, dx = 0$ fulfilled.

From this example we conclude that condition (6.2.20) of Theorem 6.2.4 is essential and in general it cannot be replaced by condition (6.2.21).

6.3 Parabolic differential equations with 'maxima'

In this section the oscillatory properties of the solutions of some classes of neutral parabolic differential equations with 'maxima' are investigated. We shall note that the problems on ordinary differential equations with 'maxima' find application in the theory of automatic regulation of various real systems [110], [132].

The necessity of investigating differential equations with 'maxima' was also underlined in the survey of Myshkis [124]. Theorems of existence and uniqueness of the solution of ordinary differential equations with 'maxima' were obtained in the works of Magomedov [111], Angelov and Bainov [3] and Konstantinov and Bainov [83]. Oscillatory and asymptotic properties of the solutions of various classes of functional differential equations with 'maxima' were investigated by Bainov and Zahariev [14] and Mishev [113], [116], [117]. In the subsequent theorems sufficient conditions for oscillation of the solutions of neutral parabolic differential equations with 'maxima' of the form

$$\frac{\partial}{\partial t} [u(x, t) + \lambda(t)u(x, \tau(t))] - [a(t) \Delta u(x, t) + b(t) \Delta u(x, \sigma(t))]$$

$$+ c(x, t, u(x, t), \max_{s \in M(t)} u(x, s)) = f(x, t), \quad (x, t) \in \Omega \times (0, \infty) \equiv G$$

$$(6.3.1)$$

are obtained, where $\Delta u(x, t) = \sum_{i=1}^{n} u_{x_i x_i}(x, t)$, $M(t) \subset [0, \infty)$ for $t \geq t_0$ and Ω is a bounded domain in $\mathbb{R}^n$ with a piecewise smooth boundary. Consider boundary conditions of the form

$$\frac{\partial u}{\partial n} + \gamma(x, t)u = g(x, t), \quad (x, t) \in \partial\Omega \times [0, \infty), \qquad (6.3.2)$$

$$u(x, t) = h(x, t), \quad (x, t) \in \partial\Omega \times [0, \infty). \qquad (6.3.3)$$

We shall say that conditions (H6.3) are satisfied if the following conditions hold:

H6.3.1 $\lambda(t) \in C^1([0, \infty); [0, \infty))$.

H6.3.2 $a(t), b(t) \in C([0, \infty); [0, \infty))$.

H6.3.3 $\tau(t) \in C^1([0, \infty); \mathbb{R})$, $\lim_{t \to \infty} \tau(t) = \infty$.

H6.3.4 $\sigma(t) \in C^2([0, \infty); \mathbb{R})$, $\lim_{t \to \infty} \sigma(t) = \infty$.

H6.3.5 $c(x, t, \beta, \eta) \in C(G \times \mathbb{R}^2; \mathbb{R})$,

$$c(x, t, \beta, \eta) \geq 0 \quad \text{for} \quad (x, t) \in G, \quad \beta \geq 0, \quad \eta \geq 0,$$

$$c(x, t, \beta, \eta) \leq 0 \quad \text{for} \quad (x, t) \in G, \quad \beta \leq 0, \quad \eta \leq 0.$$

H6.3.6 $f(x, t) \in C(G; \mathbb{R})$.

H6.3.7 $M(t)$ is a closed and bounded set for any $t \in [0, \infty)$ and $\lim_{t \to \infty} v(t) = \infty$, where $v(t) = \min_{s \in M(t)} s$.

H6.3.8 $g(x, t), h(x, t) \in C(\partial\Omega \times [0, \infty); \mathbb{R})$.

H6.3.9 $\gamma(x, t) \in C(\partial\Omega \times [0, \infty); [0, \infty))$.

Definition 6.3.1 The solution $u(x, t) \in C^2(G) \cap C^1(\bar{G})$ of problem (6.3.1), (6.3.2) ((6.3.1), (6.3.3)) is said to oscillate in the domain G if for any positive number ρ there exists a point $(x', t') \in \Omega \times [\rho, \infty)$ such that the equality $u(x', t') = 0$ holds.

With any solution $u(x, t) \in C^2(G) \cap C^1(\bar{G})$ of problem (6.3.1), (6.3.2) we associate the function

$$v(t) = \int_\Omega u(x, t) \, dx \left(\int_\Omega dx \right)^{-1}, \quad t \geq 0. \tag{6.3.4}$$

Lemma 6.3.1

Let conditions (H6.3) hold and let $u(x, t)$ be a positive (respectively negative) solution of problem (6.3.1), (6.3.2) in the domain G. Then the function $v(t)$ (respectively $v_1(t) = -v(t)$) satisfies the differential inequality

$$\frac{d}{dt}[v(t) + \lambda(t)v(\tau(t))] \leq \Phi(t), \quad t \geq t_1 \tag{6.3.5}$$

respectively

$$\frac{d}{dt}[v_1(t) + \lambda(t)v_1(\tau(t))] \leq -\Phi(t), \quad t \geq t_1 \tag{6.3.6}$$

where

$$\Phi(t) = \left[\int_\Omega f(x,t)\,dx + a(t) \int_{\partial\Omega} g(x,t)\,ds + b(t) \int_{\partial\Omega} g(x,\sigma(t))\,ds \right]$$

$$\times \left(\int_\Omega dx \right)^{-1}. \qquad (6.3.7)$$

Proof

Let $u(x,t)$ be a solution of problem (6.3.1), (6.3.2) for which $u(x,t) > 0$ for $(x,t) \in G$. Then from conditions H6.3.3, H6.3.4 and H6.3.7 it follows that there exists a number $t_1 \geq 0$ such that $u(x, \tau(t)) > 0$, $u(x, \sigma(t)) > 0$ and $\max_{s \in M(t)} u(x,s) > 0$ for $(x,t) \in \Omega \times [t_1, \infty)$. Integrate both sides of equation (6.3.1) with respect to x over the domain Ω and for $t \geq t_1$ obtain

$$\frac{d}{dt} \left[\int_\Omega u(x,t)\,dx + \lambda(t) \int_\Omega u(x,\tau(t))\,dx \right]$$

$$- \left[a(t) \int_\Omega \Delta u(x,t)\,dx + b(t) \int_\Omega \Delta u(x,\sigma(t))\,dx \right]$$

$$+ \int_\Omega c(x,t,u(x,t),\max_{s \in M(t)} u(x,s))\,dx = \int_\Omega f(x,t)\,dx. \qquad (6.3.8)$$

From Green's formula and condition H6.3.9 is follows that

$$\int_\Omega \Delta u(x,t)\,dx = \int_{\partial\Omega} \frac{\partial u}{\partial n}(x,t)\,ds = \int_{\partial\Omega} [g(x,t) - \gamma(x,t)u]\,ds$$

$$\leq \int_{\partial\Omega} g(x,t)\,ds \qquad (6.3.9)$$

$$\int_\Omega \Delta u(x, \sigma(t))\, dx = \int_{\partial\Omega} \frac{\partial u}{\partial n}(x, \sigma(t))\, ds$$

$$= \int_{\partial\Omega} [g(x, \sigma(t)) - \gamma(x, \sigma(t)) u(x, \sigma(t))]\, ds$$

$$\leqq \int_{\partial\Omega} g(x, \sigma(t))\, ds. \tag{6.3.10}$$

From condition H6.3.5 it follows that

$$\int_\Omega c(x, t, u(x, t), \max_{s \in M(t)} u(x, s))\, dx \geqq 0 \tag{6.3.11}$$

Using (6.3.9), (6.3.10) and (6.3.11), from (6.3.8) we obtain

$$\frac{d}{dt}[v(t) + \lambda(t)v(\tau(t))]$$

$$\leqq \left[\int_\Omega f(x, t)\, dx + a(t) \int_{\partial\Omega} g(x, t)\, ds + b(t) \int_{\partial\Omega} g(x, \sigma(t))\, ds \right] \left(\int_\Omega dx \right)^{-1}$$

which was to be proved.

Analogously we consider the case when $u(x, t) < 0$ for $(x, t) \in G$.

Definition 6.3.2 The solution $v(t) \in C^1([t_1, \infty); \mathbb{R})$ of the differential inequality (6.3.5) is said to be eventually positive if there exists a number $t_2 \geqq t_1$ such that the inequality $v(t) > 0$ holds for $t \geqq t_2$.

Theorem 6.3.1

Let conditions (H6.3) hold and let the differential inequalities (6.3.5) and (6.3.6) have no eventually positive solutions. Then each solution $u(x, t)$ of problem (6.3.1), (6.3.2) oscillates in the domain G.

Proof

Suppose that this is not true and let $u(x, t)$ be a solution of problem (6.3.1), (6.3.2) without a zero in the domain $G\rho = \Omega \times [\rho, \infty)$. If $u(x, t) > 0$ for $(x, t) \in G\rho$, then from Lemma 6.3.1 it follows that the function $v(t)$ defined by (6.3.4) is a positive solution of inequality (6.3.5) which contradicts the condition of the theorem. If $u(x, t) < 0$ for $(x, t) \in G\rho$, again from Lemma 6.3.1 it follows that the function $v_1(t) = -v(t)$ is a positive solution of inequality (6.3.6) which also contradicts the condition of the theorem. Thus Theorem 6.3.1 is proved.

Now we shall investigate the oscillatory properties of the solutions of problem (6.3.1), (6.3.3). Let α_0 be the least eigenvalue and let $\varphi(x)$ be the corresponding eigenfunction of problem (6.1.12), (6.1.13). It is well known that $\alpha_0 > 0$ and $\varphi(x) > 0$ for $x \in \Omega$.

With each solution $u(x, t) \in C^2(G) \cap C^1(\bar{G})$ of problem (6.3.1), (6.3.3) we associate the function

$$w(t) = \int_\Omega u(x, t)\varphi(x)\,dx \left(\int_\Omega \varphi(x)\,dx \right)^{-1}, \quad t \geq 0. \qquad (6.3.13)$$

Lemma 6.3.2

Let conditions (H6.3) hold and let $u(x, t)$ be a positive (respectively negative) solution of problem (6.3.1), (6.3.3) in the domain G. Then the function $w(t)$ (respectively $w_1(t) = -w(t)$) satisfies the differential inequality

$$\frac{d}{dt}[w(t) + \lambda(t)w(\tau(t))] + \alpha_0[a(t)w(t) + b(t)w(\sigma(t))] \leq \Phi_1(t), \quad t \geq t_1$$

$$(6.3.14)$$

respectively

$$\frac{d}{dt}[w(t) + \lambda(t)w_1(\tau(t))] + \alpha_0[a(t)w_1(t) + b(t)w_1(\sigma(t))] \leq -\Phi_1(t), \quad t \geq t_1$$

$$(6.3.15)$$

where

$$\Phi_1(t) = \left[\int_\Omega f(x, t)\varphi(x)\,dx - \int_{\partial\Omega} (a(t)h(x, t) + b(t)h(x, \sigma(t)))\frac{\partial\varphi}{\partial n}\,ds \right]$$

$$\times \left(\int_\Omega \varphi(x)\,dx \right)^{-1}. \qquad (6.3.16)$$

Proof

Let $u(x, t)$ be a solution of problem (6.3.1), (6.3.3) for which the inequality $u(x, t) > 0$ is satisfied for $(x, t) \in G$. Then from conditions H6.3.3, H6.3.4 and H6.3.7 it follows that there exists a number $t_1 \geq 0$ such that $u(x, \tau(t)) > 0$, $u(x, \sigma(t)) > 0$ and $\max_{s \in M(t)} u(x, s) > 0$ for $(x, t) \in \Omega \times [t_1, \infty)$. Multiply both sides of equation (6.3.1) by the function $\varphi(x)$ and integrate with respect to x over the domain Ω. For $t \geq t_1$ we obtain

$$\frac{d}{dt}\left[\int_\Omega u(x, t)\varphi(x)\,dx + \lambda(t)\int_\Omega u(x, \tau(t))\varphi(x)\,dx\right]$$

$$-\left[a(t)\int_\Omega \Delta u(x, t)\varphi(x)\,dx + b(t)\int_\Omega \Delta u(x, \sigma(t))\varphi(x)\,dx\right]$$

$$+\int_\Omega c\left(x, t, u(x, t), \max_{s \in M(t)} u(x, s)\right)\varphi(x)\,dx = \int_\Omega f(x, t)\varphi(x)\,dx. \quad (6.3.17)$$

From Green's formula it follows that

$$\int_\Omega \Delta u(x, t)\varphi(x)\,dx = -\int_{\partial\Omega} u(x, t)\frac{\partial\varphi}{\partial n}\,ds + \int_\Omega u(x, t)\,\Delta\varphi(x)\,dx$$

$$= -\int_{\partial\Omega} h(x, t)\frac{\partial\varphi}{\partial n}\,ds - \alpha_0\int_\Omega u(x, t)\varphi(x)\,dx,$$

$$(6.3.18)$$

$$\int_\Omega \Delta u(x, \sigma(t))\varphi(x)\,dx = -\int_{\partial\Omega} u(x, \sigma(t))\frac{\partial\varphi}{\partial n}\,ds + \int_\Omega u(x, \sigma(t))\,\Delta\varphi(x)\,dx$$

$$= -\int_{\partial\Omega} h(x, \sigma(t))\frac{\partial\varphi}{\partial n}\,ds - \alpha_0\int_\Omega u(x, \sigma(t))\varphi(x)\,dx$$

$$(6.3.19)$$

where α_0 is the least eigenvalue of problem (6.3.12). From condition H6.3.5 it follows that

$$\int_{\Omega} c(x, t, u(x, t), \max_{s \in M(t)} u(x, s))\varphi(x)\, dx \geq 0. \qquad (6.3.20)$$

Using (6.3.18), (6.3.19) and (6.3.20), from (6.3.17) we obtain

$$\frac{d}{dt}[w(t) + \lambda(t)w(\tau(t))] + \alpha_0[a(t)w(t) + b(t)w(\sigma(t))]$$

$$\leq \left[\int_{\Omega} f(x, t)\varphi(x)\, dx - a(t) \int_{\partial\Omega} h(x, t)\frac{\partial\varphi}{\partial n}\, ds - b(t) \int_{\partial\Omega} h(x, \sigma(t))\frac{\partial\varphi}{\partial n}\, ds\right]$$

$$\times \left(\int_{\Omega} \varphi(x)\, dx\right)^{-1}$$

which was to be proved. Analogously we consider the case when $u(x, t) < 0$ for $(x, t) \in G$.

By means of Lemma 6.3.2 the following theorem is proved.

Theorem 6.3.2

Let conditions (H6.3) hold and let the differential inequalities (6.3.14) and (6.3.15) have no eventually positive solutions. Then each solution $u(x, t)$ of problem (6.3.1), (6.3.3) oscillate in the domain G.

From Theorem 6.3.1 and Theorem 6.3.2 it follows that the finding of sufficient conditions for oscillation of the solutions of equation (6.3.1) is reduced to the investigation of the oscillatory properties of neutral differential inequalities of the form

$$\frac{d}{dt}[y(t) + \lambda(t)y(\tau(t))] + p(t)y(t) + q(t)y(\sigma(t)) \leq F(t),$$

$$(6.3.31)$$

$$t \geq t_0.$$

Assume the following conditions fulfilled:

H6.3.10 $\lambda(t) \in C^1([t_0, \infty); [0, \infty))$.

H6.3.11 $\tau(t) \in C^1([t_0, \infty); \mathbb{R})$, $\sigma(t) \in C([t_0, \infty); \mathbb{R})$, $\lim_{t \to \infty} \tau(t) = \lim_{t \to \infty} \sigma(t) = \infty$.

H6.3.12 $p(t), q(t) \in C([t_0, \infty); [0, \infty))$.
H6.3.13 $F(t) \in C([t_0, \infty); \mathbb{R})$.

Theorem 6.3.3

Let conditions H6.3.10–H6.3.13 hold in addition to the condition

$$\lim_{t \to \infty} \inf \int_{t_1}^{t} F(s)\,ds = -\infty \qquad (6.3.22)$$

for any number $t_1 \geq t_0$. Then the differential inequality (6.3.21) has no eventually positive solutions.

Proof

Suppose that this is not true and let $y(t)$ be an eventually positive solution of (6.3.21). From condition H6.3.11 it follows that there exists a number $t_1 \geq t_0$ such that $y(t) > 0$, $y(\tau(t)) > 0$ and $y(\sigma(t)) > 0$ for $t \geq t_1$. Then using condition H6.3.12 from (6.3.21) we obtain

$$\frac{d}{dt}[y(t) + \lambda(t)y(\tau(t))] \leq F(t).$$

Integrate the last inequality over the interval $[t_1, t]$ and obtain

$$y(t) + \lambda(t)y(\tau(t)) \leq C_1 + \int_{t_1}^{t} F(s)\,ds.$$

Then from condition (6.3.22) it follows that

$$\lim_{t \to \infty} \inf [y(t) + \lambda(t)y(\tau(t))] = -\infty. \qquad (6.3.23)$$

On the other hand, using condition H6.3.10 and the eventual positivity of the solution $y(t)$, we obtain

$$\lim_{t \to \infty} \inf [y(t) + \lambda(t)y(\tau(t))] \geq 0$$

which contradicts equality (6.3.23). This completes the proof of Theorem 6.3.3.

A corollary of Theorem 6.3.1 and Theorem 6.3.3 is the following sufficient condition for oscillation of the solutions of problem (6.3.1), (6.3.2).

Theorem 6.3.4

Let conditions (H6.3) hold in addition to the conditions

$$\liminf_{t \to \infty} \int_{t_0}^{t} \Phi(s)\, ds = -\infty, \tag{6.3.24}$$

$$\limsup_{t \to \infty} \int_{t_0}^{t} \Phi(s)\, ds = +\infty \tag{6.3.25}$$

for any sufficiently large number t_0, where the function $\Phi(t)$ is defined by (6.3.7).

Then each solution $u(x, t)$ of problem (6.3.1), (6.3.2) oscillates in G.

A corollary of Theorem 6.3.2 and Theorem 6.3.3 is the following sufficient condition for oscillation of the solutions of problem (6.3.1), (6.3.3).

Theorem 6.3.5

Let conditions (H6.3) hold in addition to the conditions

$$\liminf_{t \to \infty} \int_{t_0}^{t} \Phi_1(s)\, ds = -\infty, \tag{6.3.26}$$

$$\limsup_{t \to \infty} \int_{t_0}^{t} \Phi_1(s)\, ds = +\infty \tag{6.3.27}$$

for any sufficiently large number t_0, where the function $\Phi_1(t)$ is defined by (6.3.16).

Then each solution $u(x, t)$ of problem (6.3.1), (6.3.3) oscillates in G.

6.4 Linear hyperbolic differential equations

The development of oscillation theory for linear hyperbolic differential equations began in 1969 with the work of Kreith [90]. The works of Kreith [85], [88], [89], Kreith and Pagan [93], etc. were devoted to the further investigation of the oscillatory properties of the solutions of these classes of equations. The oscillatory properties of the solutions of linear hyperbolic equations with a deviating argument were investigated in the paper of Georgiou and Kreith [36].

In this section sufficient conditions for oscillation of the solutions of linear hyperbolic equations of neutral type of the form

$$\frac{\partial^2}{\partial t^2}\left[u(x,t) + \sum_{i=1}^{m} \lambda_i(t)u(x,t-\tau_i)\right] - \left[\Delta u(x,t) + \sum_{i=1}^{s} \mu_i(t)\,\Delta u(x,t-\rho_i)\right]$$

$$+ p(x,t)u(x,t) + \sum_{i=1}^{k} p_i(x,t)u(x,t-\sigma_i) = 0, \quad (x,t)\in\Omega \times (0,\infty) \equiv G$$

$$(6.4.1)$$

are obtained, where Ω is a bounded domain in $\mathbb{R}^n$ with a piecewise smooth boundary, $\Delta u = \sum_{i=1}^{n} u_{x_i x_i}$, $\tau_i, \rho_i, \sigma_i = \text{const.} > 0$.

Consider boundary conditions of the form

$$\frac{\partial u}{\partial n} + \gamma(x,t)u = 0, \quad (x,t)\in\partial\Omega \times [0,\infty), \qquad (6.4.2)$$

$$u = 0, \quad (x,t)\in\partial\Omega \times [0,\infty). \qquad (6.4.3)$$

We shall say that conditions (H6.4) are met if the following conditions hold:

H6.4.1 $\lambda_i(t)\in C^2([0,\infty);[0,\infty))$, $i = 1, 2, \ldots, m$.
H6.4.2 $\mu_i(t)\in C([0,\infty);[0,\infty))$, $i = 1, 2, \ldots, s$.
H6.4.3 $p(x,t), p_i(x,t)\in C(\bar{G};[0,\infty))$, $i = 1, 2, \ldots, k$.
H6.4.4 $\gamma(x,t)\in C(\partial\Omega \times [0,\infty);[0,\infty))$.

Definition 6.4.1   The solution $u(x,t)\in C^2(G)\cap C^1(\bar{G})$ of problem (6.4.1), (6.4.2) ((6.4.1), (6.4.3)) is said to oscillate in the domain $G = \Omega \times (0,\infty)$ if for any positive number μ there exists a point $(x_0,t_0)\in\Omega \times [\mu,\infty)$ such that the equality $u(x_0,t_0) = 0$ should hold.

In the subsequent theorems sufficient conditions for oscillation of the solutions of problems (6.4.1), (6.4.2) and (6.4.1), (6.4.3) in the domain G are obtained.

Introduce the following notation:

$$P(t) = \min\{p(x, t): x \in \bar{\Omega}\},$$

$$P_i(t) = \min\{p_i(x, t): x \in \bar{\Omega}\}, \quad i = 1, 2, \ldots, k.$$

(6.4.4)

With each solution $u(x, t) \in C^2(G) \cap C^1(\bar{G})$ of problem (6.4.1), (6.4.2) we associate the function

$$v(t) = \int_{\Omega} u(x, t)\, dx, \quad t \geq 0.$$

(6.4.5)

Lemma 6.4.1

Let conditions (H6.4) hold and let $u(x, t)$ be a positive solution of problem (6.4.1), (6.4.2) in the domain G. Then the function $v(t)$ defined by (6.4.5) satisfies the differential inequality

$$\frac{d^2}{dt^2}\left[v(t) + \sum_{i=1}^{m} \lambda_i(t) v(t - \tau_i)\right] + P(t) v(t) + \sum_{i=1}^{k} P_i(t) v(t - \sigma_i) \leq 0, \quad t \geq t_0$$

(6.4.6)

where t_0 is a sufficiently large positive number.

Proof

Let $u(x, t)$ be a positive solution of problem (6.4.1), (6.4.2) in the domain G and $t_0 = \max\{\tau_1, \ldots, \tau_m, \rho_1, \ldots, \rho_s, \sigma_1, \ldots, \sigma_k\}$. Then $u(x, t - \tau_i) > 0$, $u(x, t - \rho_i) > 0$ and $u(x, t - \sigma_i) > 0$ for $(x, t) \in \Omega \times (t_0, \infty)$. Integrate both sides of equation (6.4.1) with respect to x over the domain Ω and for $t \geq t_0$ obtain

$$\frac{d^2}{dt^2}\left[\int_{\Omega} u(x, t)\, dx + \sum_{i=1}^{m} \lambda_i(t) \int_{\Omega} u(x, t - \tau_i)\, dx\right]$$

$$- \left[\int_{\Omega} \Delta u(x, t)\, dx + \sum_{i=1}^{s} \mu_i(t) \int_{\Omega} \Delta u(x, t - \rho_i)\, dx\right]$$

$$+ \int_{\Omega} p(x, t) u(x, t)\, dx + \sum_{i=1}^{k} \int_{\Omega} p_i(x, t) u(x, t - \sigma_i)\, dx = 0. \quad (6.4.7)$$

From Green's formula and conditions H6.4.4 it follows that

$$\int_{\Omega} \Delta u(x, t)\, dx = \int_{\partial\Omega} \frac{\partial u}{\partial n}\, ds = -\int_{\partial\Omega} \gamma(x, t)u\, ds \leq 0 \qquad (6.4.8)$$

$$\int_{\Omega} \Delta u(x, t - \rho_i)\, dx = \int_{\partial\Omega} \frac{\partial u}{\partial n}(x, t - \rho_i)\, ds$$

$$= -\int_{\partial\Omega} \gamma(x, t - \rho_i)u(x, t - \rho_i)\, ds \leq 0. \qquad (6.4.9)$$

Moreover, from (6.4.4) it follows that

$$\int_{\Omega} p(x, t)u(x, t)\, dx \geq P(t) \int_{\Omega} u(x, t)\, dx = P(t)v(t) \qquad (6.4.10)$$

$$\int_{\Omega} p_i(x, t)u(x, t - \sigma_i)\, dx \geq P_i(t) \int_{\Omega} u(x, t - \sigma_i)\, dx = P_i(t)v(t - \sigma_i).$$

$$(6.4.11)$$

Using (6.4.8)–(6.4.11) and conditions H6.4.2, from (6.4.7) we obtain

$$\frac{d^2}{dt^2}\left[v(t) + \sum_{i=1}^{m} \lambda_i(t)v(t - \tau_i) \right] \leq -P(t)v(t) - \sum_{i=1}^{k} P_i(t)v(t - \sigma_i)$$

which completes the proof of Lemma 6.4.1.

Definition 6.4.2   The solution $v(t) \in C^2([t_0, \infty); \mathbb{R})$ of the differential inequality (6.4.6) is said to be eventually positive (negative) if there exists a sufficiently large positive number t_1 such that the inequality $v(t) > 0$ $(v(t) < 0)$ holds for $t \geq t_1$.

Theorem 6.4.1

Let conditions (H6.4) hold and let the differential inequality (6.4.6) have no eventually positive solutions. Then each solution $u(x, t)$ of problem (6.4.1), (6.4.2) oscillates in the domain G.

Proof

Let $\mu > 0$ be an arbitrary number. Suppose that the assertion is not true and let $u(x, t)$ be a solution of problem (6.4.1), (6.4.2) without a zero in the domain $G_\mu = \Omega \times [\mu, \infty)$. If $u(x, t) > 0$ for $(x, t) \in G_\mu$, then from Lemma 6.4.1 it follows that the function $v(t)$ defined by (6.4.5) is a positive solution of inequality (6.4.6) for $t \geq t_0 + \mu$ which contradicts the condition of the theorem. If $u(x, t) < 0$ for $(x, t) \in G_\mu$, then $-u(x, t)$ is a positive solution of problem (6.4.1), (6.4.2). From Lemma 6.4.1 it follows that the function $v_1(t) = -v(t) = -\int_\Omega u(x, t) \, dx$ is a positive solution of inequality (6.4.6) for $t \geq t_0 + \mu$ which also contradicts the condition of the theorem.

Now we shall investigate the oscillatory properties of the solutions of problem (6.4.1), (6.4.3). Let α_0 be the least eigenvalue of problem (6.1.12), (6.1.13) and let $\varphi(x)$ be the corresponding eigenfunction. It is well known [150] that $\alpha_0 > 0$ and $\varphi(x)$ can be chosen so that $\varphi(x) > 0$ for $x \in \Omega$.

With each solution $u(x, t) \in C^2(G) \cap C^1(\bar{G})$ of problem (6.4.1), (6.4.3), we associate the function

$$w(t) = \int_\Omega u(x, t) \varphi(x) \, dx, \quad t \geq 0. \tag{6.4.12}$$

Lemma 6.4.2

Let conditions H6.4.1–H6.4.3 hold and let $u(x, t)$ be a positive solution of problem (6.4.1), (6.4.3) in the domain G. Then the function $w(t)$ defined by (6.4.12) satisfies the differential inequality

$$\frac{d^2}{dt^2}\left[w(t) + \sum_{i=1}^{m} \lambda_i(t) w(t - \tau_i) \right] + \alpha_0 \left[w(t) + \sum_{i=1}^{s} \mu_i(t) w(t - \rho_i) \right]$$

$$+ P(t)w(t) + \sum_{i=1}^{k} P_i(t) w(t - \sigma_i) \leq 0, \quad t \geq t_0 \tag{6.4.13}$$

where t_0 is a sufficiently large positive number.

Proof

Let $u(x, t)$ be a positive solution of problem (6.4.1), (6.4.3) in the domain G and $t_0 = \max\{\tau_1, \ldots, \tau_m, \rho_1, \ldots, \rho_s, \sigma_1, \ldots, \sigma_k\}$. Then $u(x, t - \tau_i) > 0$, $u(x, t - \rho_i) > 0$ and $u(x, t - \sigma_i) > 0$ for $(x, t) \in \Omega \times (t_0, \infty)$. Multiply both sides of equation (6.4.1) by the eigenfunction $\varphi(x)$ and integrate with respect to x over the domain Ω. For $t \geq t_0$ we obtain

$$\frac{d^2}{dt^2}\left[\int_\Omega u(x,t)\varphi(x)\,dx + \sum_{i=1}^m \lambda_i(t)\int_\Omega u(x,t-\tau_i)\varphi(x)\,dx\right]$$

$$-\left[\int_\Omega \Delta u(x,t)\varphi(x)\,dx + \sum_{i=1}^s \mu_i(t)\int_\Omega \Delta u(x,t-\rho_i)\varphi(x)\,dx\right]$$

$$+\int_\Omega p(x,t)u(x,t)\varphi(x)\,dx + \sum_{i=1}^k \int_\Omega p_i(x,t)u(x,t-\sigma_i)\varphi(x)\,dx = 0.$$

$$(6.4.14)$$

From Green's formula it follows that

$$\int_\Omega \Delta u(x,t)\varphi(x)\,dx = \int_\Omega u(x,t)\,\Delta\varphi(x)\,dx = -\alpha_0 \int_\Omega u(x,t)\varphi(x)\,dx$$

$$= -\alpha_0 w(t), \qquad (6.4.15)$$

$$\int_\Omega \Delta u(x,t-\rho_i)\varphi(x)\,dx = \int_\Omega u(x,t-\rho_i)\,\Delta\varphi(x)\,dx$$

$$= -\alpha_0 \int_\Omega u(x,t-\rho_i)\varphi(x)\,dx$$

$$= -\alpha_0 w(t-\rho_i). \qquad (6.4.16)$$

Moreover, from (6.4.4) it follows that

$$\int_\Omega p(x,t)u(x,t)\varphi(x)\,dx \geq P(t)\int_\Omega u(x,t)\varphi(x)\,dx = P(t)w(t) \qquad (6.4.17)$$

$$\int_\Omega p_i(x,t)u(x,t-\sigma_i)\varphi(x)\,dx \geq P_i(t)\int_\Omega u(x,t-\sigma_i)\varphi(x)\,dx$$

$$= P_i(t)w(t-\sigma_i). \qquad (6.4.18)$$

Using (6.4.15)–(6.4.18) and conditions H6.4.2 from (6.4.14) we obtain

$$\frac{\mathrm{d}^2}{\mathrm{d}t^2}\left[w(t) + \sum_{i=1}^{m} \lambda_i(t)w(t - \tau_i)\right] \leq -\alpha_0\left[w(t) + \sum_{i=1}^{s} \mu(t)w(t - \rho_i)\right]$$

$$- P(t)w(t) - \sum_{i=1}^{k} P_i(t)w(t - \sigma_i).$$

This completes the proof of Lemma 6.4.2.

Analogously to Theorem 6.4.1 the following theorem is proved.

Theorem 6.4.2

Let conditions H6.4.1–H6.4.3 hold and let the differential inequality (6.4.13) have no eventually positive solutions. Then each solution $u(x, t)$ of problem (6.4.1), (6.4.3) oscillates in the domain G.

From the theorems proved above in this section it follows that the finding of sufficient conditions for oscillation of the solutions of equation (6.4.1) in the domain G is reduced to the investigation of the oscillatory properties of neutral differential inequalities of the form

$$\frac{\mathrm{d}^2}{\mathrm{d}t^2}\left[x(t) + \sum_{i=1}^{m} \lambda_i(t)x(t - \tau_i)\right] + q(t)x(t) + \sum_{i=1}^{k} q_i(t)x(t - \sigma_i) \leq 0, \quad t \geq t_0$$

(6.4.19)

$$\frac{\mathrm{d}^2}{\mathrm{d}t^2}\left[x(t) + \sum_{i=1}^{m} \lambda_i(t)x(t - \tau_i)\right] + q(t)x(t) + \sum_{i=1}^{k} q_i(t)x(t - \sigma_i) \geq 0, \quad t \geq t_0.$$

(6.4.20)

Together with (6.4.19) and (6.4.20) we shall consider the neutral differential equation

$$\frac{\mathrm{d}^2}{\mathrm{d}t^2}\left[x(t) + \sum_{i=1}^{m} \lambda_i(t)x(t - \tau_i)\right] + q(t)x(t) + \sum_{i=1}^{k} q_i(t)x(t - \sigma_i) = 0,$$

$$t \geq t_0.$$

(6.4.21)

Assume the following conditions are fulfilled:

H6.4.5 $\lambda_i(t) \in C^2([t_0, \infty); [0, \infty))$, $i = 1, 2, \ldots, m$.

H6.4.6 $q(t), q_i(t) \in C([t_0, \infty); [0, \infty))$, $i = 1, 2, \ldots, k$.

Theorem 6.4.3

Let conditions H6.4.5–H6.4.6 hold in addition to the following conditions

$$\sum_{i=1}^{m} \lambda_i(t) \le 1 \quad \text{for} \quad t \ge t_0 \tag{6.4.22}$$

$$\int_{t_0}^{\infty} q_v(t) \left[1 - \sum_{i=1}^{m} \lambda_i(t - \sigma_v) \right] dt = \infty \tag{6.4.23}$$

for at least one number $v \in \{1, 2, \ldots, k\}$.
  Then:
  (i)   the differential inequality (6.4.19) has no eventually positive solutions;
  (ii) the differential inequality (6.4.20) has no eventually negative solutions;
  (iii) all solutions of the differential equation (6.4.21) oscillate.

Proof

(i) Suppose that there exists an eventually positive solution $x(t)$ of the differential inequality (6.4.19). Hence $x(t) > 0$ for $t \ge t_1$, where $t_1 \ge t_0$.
  Introduce the notation

$$z(t) = x(t) + \sum_{i=1}^{m} \lambda_i(t) x(t - \tau_i), \quad t \ge t_1 + \tau,$$

$$\tau = \max\{\tau_1, \ldots, \tau_m\}, \quad \sigma = \max\{\sigma_1, \ldots, \sigma_k\}. \tag{6.4.24}$$

From condition H6.4.5 it follows that

$$z(t) > 0 \quad \text{for} \quad t \ge t_1 + \tau. \tag{6.4.25}$$

Using condition H6.4.6, from the differential inequality (6.4.19) for $t \ge t_2 = t_1 + \tau + \sigma$ we obtain

$$z''(t) \le -q(t) x(t) - \sum_{i=1}^{k} q_i(t) x(t - \sigma_i) \le 0.$$

Hence the function $z'(t)$ is monotone decreasing in the interval $[t_2, \infty)$. We shall prove the inequality

$$z'(t) \ge 0 \quad \text{for} \quad t \ge t_2. \tag{6.4.26}$$

Suppose that there exists a number $t_3 \geq t_2$ such that $z'(t_3) = -C < 0$. Then for any $t \geq t_3$ the following inequality holds

$$z'(t) \leq z'(t_3) = -C.$$

Integrate the last inequality over the interval $[t_3, t]$, $t > t_3$ and obtain

$$z(t) \leq z(t_3) - C(t - t_3).$$

Hence $\lim_{t \to \infty} \sup z(t) \leq 0$ which contradicts (6.4.25). Thus inequality (6.4.26) is proved.

Using condition H6.4.6, from (6.4.19) for $t \geq t_2$ we obtain

$$z''(t) + q_v(t)x(t - \sigma_v) \leq 0 \qquad (6.4.27)$$

where $v \in \{1, 2, \ldots, k\}$ and (6.4.23) holds. From (6.4.24) and (6.4.27) it follows that

$$z''(t) + q_v(t)\left[z(t - \sigma_v) - \sum_{i=1}^{m} \lambda_i(t - \sigma_v)x(t - \sigma_v - \tau_i) \right] \leq 0.$$

Using that $z(t) \geq x(t)$ for $t \geq t_2$ and (6.4.26), from the last inequality we obtain

$$z''(t) + q_v(t)\left[1 - \sum_{i=1}^{m} \lambda_i(t - \sigma_v) \right] z(t - \sigma_v) \leq 0. \qquad (6.4.28)$$

Integrating inequality (6.4.28) over the interval $[t_2, t]$, $t > t_2$ and using condition (6.4.22) and (6.4.26) we obtain

$$z'(t) - z'(t_2) + z(t_2 - \sigma_v) \int_{t_2}^{t} q_v(t)\left[1 - \sum_{i=1}^{m} \lambda_i(t - \sigma_v) \right] dt \leq 0.$$

For $t \to \infty$ from the above inequality it follows that

$$\int_{t_2}^{\infty} q_v(t)\left[1 - \sum_{i=1}^{m} \lambda_i(t - \sigma_v) \right] dt < \infty$$

which contradicts condition (6.4.23).

Thus assertion (i) of Theorem 6.4.3 is proved.

(*ii*) The proof follows from the fact that $x(t)$ is an eventually negative solution of the differential inequality (6.4.20), then $-x(t)$ is an eventually positive solution of the differential inequality (6.4.19).

(*iii*) From (*i*) and (*ii*) it follows that (6.4.21) has no eventually positive and eventually negative solutions. Hence all solutions of the differential equation (6.4.21) oscillate.

A corollary of Theorem 6.4.1 and Theorem 6.4.3 is the following sufficient condition for oscillation of the solutions of problem (6.4.1), (6.4.2).

Theorem 6.4.4

Let conditions (H6.4), condition (6.4.22) and the following condition hold:

$$\int_{t_0}^{\infty} P_v(t)\left[1 - \sum_{i=1}^{m} \lambda_i(t - \sigma_v)\right] dt = \infty \qquad (6.4.29)$$

for at least one number $v \in \{1, 2, \ldots, k\}$.
Then each solution $u(x, t)$ of problem (6.4.1), (6.4.2) oscillates in G.

A corollary of Theorem 6.4.2 and Theorem 6.4.3 is the following sufficient condition for oscillation of the solutions of problem (6.4.1), (6.4.3).

Theorem 6.4.5

Let conditions (H6.4), condition (6.4.22) and the following condition hold:

$$\int_{t_0}^{\infty} \mu_v(t)\left[1 - \sum_{i=1}^{m} \lambda_i(t - \rho_v)\right] dt = \infty \qquad (6.4.30)$$

for at least one number $v \in \{1, 2, \ldots, s\}$, or condition (6.4.29).
Then each solution $u(x, t)$ of problem (6.4.1), (6.4.3) oscillates in G.

Example 6.4.1 Consider the equation

$$u_{tt} + \tfrac{1}{2}u_{tt}(x, t - \pi) - [u_{xx} + 2u_{xx}(x, t - \pi)] + \tfrac{3}{2}u + u(x, t - \pi) = 0,$$

$$(x, t) \in (0, \pi) \times (0, \infty) \equiv G \qquad (6.4.31)$$

with boundary condition

$$u_x(0, t) = u_x(\pi, t) = 0, \quad t \geq 0. \tag{6.4.32}$$

A straightforward verification shows that the functions

$$\lambda_1(t) \equiv \tfrac{1}{2}, \quad \mu_1(t) \equiv 2, \quad p(x, t) \equiv \tfrac{5}{2}, \quad p_1(x, t) \equiv 1, \quad \gamma(x, t) \equiv 0$$

satisfy all conditions of Theorem 6.4.4. Hence all solutions of problem (6.4.31), (6.4.32) oscillate in the domain G. For instance, the function $u(x, t) = \cos x \cos t$ is such a solution.

Example 6.4.2 Consider the equation

$$u_{tt} + e^{-\pi} u_{tt}(x, t - \pi) - [u_{xx} + 3e^{-\pi} u_{xx}(x, t - \pi)] + 2u = 0,$$

$$\tag{6.4.33}$$

$$(x, t) \in (0, \pi) \times (0, \infty) \equiv G$$

with boundary condition

$$u(0, t) = u(\pi, t) = 0, \quad t \geq 0. \tag{6.4.34}$$

A straightforward verification shows that the functions

$$\lambda_1(t) = e^{-\pi}, \quad \mu_1(t) = 3e^{-\pi}, \quad p(x, t) \equiv 2, \quad p_1(x, t) \equiv 0$$

satisfy all conditions of Theorem 6.4.5. Hence all solutions of problem (6.4.33), (6.4.34) oscillate in the domain G. For instance, the function $u(x, t) = e^{-t} \sin x \cos t$ is such a solution.

6.5 Nonlinear hyperbolic differential equations

The development of the oscillation theory for nonlinear hyperbolic differential equations began in 1979 with the work of Yoshida [152]. The works of Kreith *et al* [91], Hsiang and Kwong [69], [70] and Yoshida [155], [157] were devoted to the further investigation of the oscillatory properties of the solutions of these classes of equations. Oscillatory properties of the solutions of nonlinear hyperbolic equations with a translated argument were investigated in the work of Georgiou and Kreith [36] and of nonlinear hyperbolic equations of neutral type in the papers of Mishev and Bainov [119], [121], [122] and Yoshida [154].

In this section sufficient conditions are obtained for oscillation of the solutions of neutral nonlinear hyperbolic equations of the form

$$\frac{\partial^2}{\partial t^2}[u(x,t)+\lambda(t)u(x,t-\tau)]-[\Delta u(x,t)+\mu(t)\Delta u(x,t-\sigma)]$$

$$+c(x,t,u)=f(x,t), \quad (x,t)\in\Omega\times(0,\infty)\equiv G \qquad (6.5.1)$$

where $\tau, \sigma = \text{const.} > 0$, $\Delta u(x,t)=\sum_{i=1}^{n}u_{x_i x_i}(x,t)$ and Ω is a bounded domain $\mathbb{R}^n$ with a piecewise smooth boundary.

Consider boundary conditions of the form

$$\frac{\partial u}{\partial n}+\gamma(x,t)u=g(x,t), \quad (x,t)\in\partial\Omega\times[0,\infty), \qquad (6.5.2)$$

$$u=0, \quad (x,t)\in\partial\Omega\times[0,\infty). \qquad (6.5.3)$$

We shall say that conditions (H6.5) are met if the following conditions hold:

H6.5.1 $\lambda(t)\in C^2([0,\infty);[0,\infty))$,

$\mu(t)\in C([0,\infty);\mathbb{R})$.

H6.5.2 $c(x,t,u)\in C(G\times\mathbb{R};\mathbb{R})$.
H6.5.3 $c(x,t,-u)=-c(x,t,u)$, $(x,t,u)\in G\times(0,\infty)$.
H6.5.4 $c(x,t,u)\geq p(t)h(u)$, $(x,t,u)\in G\times(0,\infty)$,
where $p(t)$ is a continuous and positive function in the interval $(0,\infty)$ and $h(u)$ is a continuous, positive and convex function in the same interval $(0,\infty)$.

H6.5.5 $f(x,t)\in C(G;\mathbb{R})$.
H6.5.6 $g(x,t)\in C(\partial\Omega\times[0,\infty);\mathbb{R})$.
H6.5.7 $\gamma(x,t)\in C(\partial\Omega\times[0,\infty);[0,\infty))$.

Definition 6.5.1   The solution $u(x,t)\in C^2(G)\cap C^1(\bar{G})$ of problem (6.5.1), (6.5.2) ((6.5.1), (6.5.3)) is said to oscillate in the domain G if for any positive number μ there exists a point $(x_0,t_0)\in\Omega\times[\mu,\infty)$ such that the equality $u(x_0,t_0)=0$ holds.

In the subsequent theorems sufficient conditions for oscillation of the solutions of problems (6.5.1), (6.5.2) and (6.5.1), (6.5.3) in the domain G are obtained.

Introduce the following notation:

$$F(t) = \frac{1}{|\Omega|} \int_{\Omega} f(x, t)\, dx, \quad t \geq 0$$

$$(6.5.4)$$

$$F(t) = \frac{1}{|\Omega|} \int_{\partial\Omega} f(x, t)\, dx, \quad t \geq 0$$

where $|\Omega| = \int_{\Omega} dx$.

With each solution $u(x, t) \in C^2(G) \cap C^1(\bar{G})$ of problem (6.5.1), (6.5.2) we associate the function

$$v(t) = \frac{1}{|\Omega|} \int_{\Omega} u(x, t)\, dx, \quad t \geq 0. \qquad (6.5.5)$$

Lemma 6.5.1

Let conditions (H6.5) hold and let $u(x, t)$ be a positive solution of problem (6.5.1), (6.5.2) in the domain G. Then the function $v(t)$ defined by (6.5.5) satisfies the neutral differential inequality

$$\frac{d^2}{dt^2}[v(t) + \lambda(t)v(t - \tau)] + p(t)h(v(t)) \leq G(t) + \mu(t)G(t - \sigma) + F(t)$$

$$(6.5.6)$$

$t \geq t_0$, where t_0 is a sufficiently large positive number.

Proof

Let $u(x, t)$ be a positive solution in the domain G of problem (6.5.1), (6.5.2) and $t_0 = \max\{\tau, \sigma\}$. Then $u(x, t - \tau) > 0$ and $u(x, t - \sigma) > 0$ for $(x, t) \in \Omega \times [t_0, \infty)$. Integrate both sides of equation (6.5.1) with respect to x over the domain Ω and obtain for $t \geq t_0$

$$\frac{d^2}{dt^2}\left[\int_{\Omega} u(x, t)\, dx + \lambda(t) \int_{\Omega} u(x, t - \tau)\, dx \right]$$

$$- \left[\int_{\Omega} \Delta u(x, t)\, dx + \mu(t) \int_{\Omega} \Delta u(x, t - \sigma)\, dx \right] + \int_{\Omega} \equiv c(x, t, u)\, dx$$

$$= \int_{\Omega} f(x, t)\, dx. \qquad (6.5.7)$$

Nonlinear Hyperbolic Differential Equations 237

From Green's formula and conditions H6.5.7 it follows that

$$\int_\Omega \Delta u(x,t)\,dx = \int_{\partial\Omega} \frac{\partial u}{\partial n}\,ds = \int_{\partial\Omega} [g(x,t) - \gamma(x,t)u]\,ds \le \int_{\partial\Omega} g(x,t)\,ds$$

$$(6.5.8)$$

$$\int_\Omega \Delta u(x,t-\sigma)\,dx = \int_{\partial\Omega} \frac{\partial u}{\partial n}(x,t-\sigma)\,ds$$

$$= \int_{\partial\Omega} [g(x,t-\sigma) - \gamma(x,t-\sigma)u(x,t-\sigma)]\,ds$$

$$\le \int_{\partial\Omega} g(x,t-\sigma)\,ds. \qquad (6.5.9)$$

Moreover, from condition H6.5.4 and Jensen's inequality it follows that

$$\int_\Omega c(x,t,u)\,dx \ge p(t)\int_\Omega h(u(x,t))\,dx$$

$$\ge p(t)h\left(\int_\Omega u(x,t)\,dx\left(\int dx\right)^{-1}\right)\int_\Omega dx = p(t)h(v(t))|\Omega|. \qquad (6.5.10)$$

Using (6.5.8)–(6.5.10) and conditions H6.5.1, from (6.5.7) we obtain

$$\frac{d^2}{dt^2}[v(t) + \lambda(t)v(t-\tau)] \le G(t) + \mu(t)G(t-\sigma) + F(t) - p(t)h(v(t))$$

with which Lemma 6.5.1 is proved.

Theorem 6.5.1

Let conditions (H6.5) hold and let the neutral differential inequalities

$$\frac{d^2}{dt^2}[v(t) + \lambda(t)v(t-\tau)] + p(t)h(v(t)) \le G(t) + \mu(t)G(t-\sigma) + F(t),$$

$$t \ge t_0 \quad (6.5.11)$$

$$\frac{d^2}{dt^2}[v(t) + \lambda(t)v(t - \tau)] + p(t)h(v(t)) \leqq -[G(t) + \mu(t)G(t - \sigma) + F(t)],$$

$$t \geq t_0 \quad (6.5.12)$$

have no eventually positive solutions. Then each solution $u(x, t)$ of problem (6.5.1), (6.5.2) oscillates in the domain G.

Proof

Let $\mu > 0$ be an arbitrary number. Suppose that the assertion is not true and let $u(x, t)$ be a solution of problem (6.5.1), (6.5.2) without a zero in the domain $G_\mu = \Omega \times [\mu, \infty)$. If $u(x, t) > 0$ for $(x, t) \in G_\mu$, then from Lemma 6.5.1 it follows that the function $v(t)$ defined by (6.5.5) is a positive solution of inequality (6.5.11) for $t \geq t_0 + \mu$ which contradicts the condition of the theorem. If $u(x, t) < 0$ for $(x, t) \in G_\mu$, then the function $-u(x, t)$ is a positive solution of the problem

$$\begin{cases} \dfrac{\partial^2}{\partial t^2}[u + \lambda(t)u(x, t - \tau)] - [\Delta u + \mu(t)\,\Delta u(x, t - \sigma)] + c(x, t, u) = -f(x, t), \\ \qquad\qquad\qquad\qquad\qquad\qquad\qquad\qquad\qquad\qquad\qquad (x, t) \in G \\ \\ \dfrac{\partial u}{\partial n} + \gamma(x, t)u = -g(x, t), \quad (x, t) \in \partial\Omega \times [0, \infty). \end{cases}$$

From Lemma 6.5.1 it follows that the function

$$\frac{1}{|\Omega|}\int_\Omega [-u(x, t)]\,dx$$

is a positive solution of inequality (6.5.12) for $t \geq t_0 + \mu$ which also contradicts the condition of the theorem. Thus Theorem 6.5.1 is proved.

Now we shall investigate the oscillatory properties of the solutions of problem (6.5.1), (6.5.3). Let α_0 be the least eigenvalue of problem (6.1.12), (6.1.13) and let $\varphi(x)$ be the corresponding eigenfunction. It is well known [150] that $\alpha_0 > 0$ and $\varphi(x)$ can be chosen so that $\varphi(x) > 0$, $x \in \Omega$.

With each solution $u(x, t) \in C^2(G) \cap C^1(\bar{G})$ of problem (6.5.1), (6.5.3) we associate the function

$$w(t) = \int_\Omega u(x, t)\varphi(x)\,dx \left(\int_\Omega \varphi(x)\,dx\right)^{-1}, \quad t \geq 0 \quad (6.5.13)$$

Lemma 6.5.2

Let conditions H6.5.1–H6.5.6 hold and let $u(x, t)$ be a positive solution in the domain G of problem (6.5.1), (6.5.3). Then the function $w(t)$ defined by (6.5.13) satisfies the neutral differential inequality

$$\frac{d^2}{dt^2} [w(t) + \lambda(t)w(t - \tau)] + \alpha_0 w(t) + \alpha_0 \mu(t)w(t - \sigma) + p(t)h(w(t))$$

$$\leq \int_\Omega f(x, t)\varphi(x) \, dx \left(\int_\Omega \varphi(x) \, dx \right)^{-1}, \quad t \geq t_0 \qquad (6.5.14)$$

where t_0 is a sufficiently large positive number.

Proof

Let $u(x, t)$ be a positive solution in the domain G of problem (6.5.1), (6.5.3) and $t_0 = \max\{\tau, \sigma\}$. Then $u(x, t - \tau) > 0$ and $u(x, t - \sigma) > 0$ for $(x, t) \in \Omega \times (t_0, \infty)$. Multiply both sides of equation (6.5.1) by the eigenfunction $\varphi(x)$ and integrate with respect to x over the domain Ω. For $t \geq t_0$ we obtain:

$$\frac{d^2}{dt^2} \left[\int_\Omega u(x, t)\varphi(x) \, dx + \lambda(t) \int_\Omega u(x, t - \tau)\varphi(x) \, dx \right]$$

$$- \left[\int_\Omega \Delta u(x, t)\varphi(x) \, dx + \mu(t) \int_\Omega \Delta u(x, t - \sigma)\varphi(x) \, dx \right]$$

$$+ \int_\Omega c(x, t, u)\varphi(x) \, dx = \int_\Omega f(x, t)\varphi(x) \, dx. \qquad (6.5.15)$$

From Green's formula it follows that

$$\int_\Omega \Delta u(x, t)\varphi(x) \, dx = \int_\Omega u(x, t) \, \Delta\varphi(x) \, dx = -\alpha_0 \int_\Omega u(x, t)\varphi(x) \, dx$$

$$= -\alpha_0 w(t) \int_\Omega \varphi(x) \, dx. \qquad (6.5.16)$$

$$\int_\Omega \Delta u(x, t - \sigma)\varphi(x)\, dx = \int_\Omega u(x, t - \sigma)\, \Delta\varphi(x)\, dx$$

$$= -\alpha_0 \int_\Omega u(x, t - \sigma)\varphi(x)\, dx$$

$$= -\alpha_0 w(t - \sigma) \int_\Omega \varphi(x)\, dx. \qquad (6.5.17)$$

Moreover, from condition H6.5.4 and Jensen's inequality it follows that

$$\int_\Omega c(x, t, u)\varphi(x)\, dx \geqq p(t) \int_\Omega h(u)\varphi(x)\, dx$$

$$\geqq p(t)h\left(\int_\Omega u(x, t)\varphi(x)\, dx \left(\int_\Omega \varphi(x)\, dx\right)^{-1}\right) \int_\Omega \varphi(x)\, dx$$

$$= p(t)h(w(t)) \int_\Omega \varphi(x)\, dx. \qquad (6.5.18)$$

Using (6.5.16)–(6.5.18) and condition H6.5.1, from (6.5.15), we obtain

$$\frac{d^2}{dt^2}[w(t) + \lambda(t)w(t - \tau)] \leqq -\alpha_0[w(t) + \mu(t)w(t - \sigma)] - p(t)h(w(t))$$

$$+ \int_\Omega f(x, t)\varphi(x)\, dx \left(\int_\Omega \varphi(x)\, dx\right)^{-1}$$

which proves Lemma 6.5.2.
Introduce the following notation

$$F_1(t) = \int_\Omega f(x, t)\varphi(x)\, dx \left(\int_\Omega \varphi(x)\, dx\right)^{-1}, \quad t \geqq 0. \qquad (6.5.19)$$

Analogously to Theorem 6.5.1 the following theorem is proved.

Theorem 6.5.2

Let conditions H6.5.1–H6.5.5 hold and let the neutral differential inequalities

$$\frac{d^2}{dt^2}[w(t) + \lambda(t)w(t-\tau)] + \alpha_0[w(t) + \mu(t)w(t-\sigma)] + p(t)h(w(t))$$

$$\leq F_1(t), \quad t \geq t_0 \qquad (6.5.20)$$

$$\frac{d^2}{dt^2}[w(t) + \lambda(t)w(t-\tau)] + \alpha_0[w(t) + \mu(t)w(t-\sigma)] + p(t)h(w(t))$$

$$\leq -F_1(t), \quad t \geq t_0 \qquad (6.5.21)$$

have no eventually positive solutions. Then each solution $u(x,t)$ of problem (6.5.1), (6.5.3) oscillates in the domain G.

From the theorems proved above in this section it follows that the finding of sufficient conditions for oscillation of the solutions of equation (6.5.1) in the domain G is reduced to the investigation of the oscillatory properties of neutral differential inequalities of the form

$$\frac{d^2}{dt^2}[x(t) + \lambda(t)x(t-\tau)] + q_0(t)x(t) + q(t)x(t-\sigma) + p(t)h(x(t))$$

$$\leq H(t), \quad t \geq t_0. \qquad (6.5.22)$$

Assume the following conditions fulfilled:
H6.5.8 $\lambda(t) \in C^2([t_0, \infty); [0, \infty))$
H6.5.9 $q_0(t), q(t) \in C([t_0, \infty); [0, \infty))$
H6.5.10 $p(t) \in C([t_0, \infty); [0, \infty))$
H6.5.11 $h(u) \in C(\mathbb{R}; \mathbb{R}), \quad h(u) > 0$ for $u > 0$
H6.5.12 $H(t) \in C([t_0, \infty); \mathbb{R})$.

Theorem 6.5.3

Let conditions H6.5.8–H6.5.12 hold in addition to the condition

$$\liminf_{t \to \infty} \frac{1}{t-t_1} \int_{t_1}^{t} (t-s)H(s)\,ds = -\infty \qquad (6.5.23)$$

for any number $t_1 \geq t_0$. Then the differential inequality (6.5.22) has no eventually positive solutions.

Proof

Suppose that this is not true and let $x(t)$ be a positive solution of inequality (6.5.22) defined in the interval $[t_1, \infty)$, where $t_1 \geq t_0$. Then, using conditions H6.5.9–H6.5.11, for $t \geq t_2$, where $t_2 \geq t_1 + \max\{\sigma, \tau\}$, we obtain

$$\frac{d^2}{dt^2}[x(t) + \lambda(t)x(t - \tau)] \leq H(t) - q_0(t)x(t) - q(t)x(t - \sigma)$$

$$- p(t)h(x(t)) \leq H(t)$$

Integrating the above inequality twice over the interval $[t_2, t], t > t_2$ we obtain

$$x(t) + \lambda(t)x(t - \tau) \leq C_1 + C_2(t - t_2) + \int_{t_2}^{t}\left[\int_{t_2}^{\rho} H(s)\,ds\right]d\rho$$

where $C_1, C_2 = $ const. Since

$$\int_{t_2}^{t}\left[\int_{t_2}^{\rho} H(s)\,ds\right]d\rho = \int_{t_2}^{t}(t - s)H(s)\,ds.$$

Dividing both sides of the last inequality by $t - t_2 > 0$, we obtain

$$\frac{x(t) + \lambda(t)x(t - \tau)}{t - t_2} \leq \frac{C_1}{t - t_2} + C_2 + \frac{1}{t - t_2}\int_{t_2}^{t}(t - s)H(s)\,ds. \qquad (6.5.24)$$

Then for $t \to \infty$ from (6.5.24), using condition (6.5.23), it follows that

$$\lim_{t \to \infty} \inf \frac{x(t) + \lambda(t)x(t - \tau)}{t - t_2} = -\infty. \qquad (6.5.25)$$

On the other hand, using condition H6.5.8 and the fact that $x(t) > 0$, $x(t - \tau) > 0$ for $t \geq t_2$, we obtain that

$$\lim_{t \to \infty} \inf \frac{1}{t - t_2}[x(t) + \lambda(t)x(t - \tau)] \geq 0$$

which contradicts equality (6.5.25).
    This completes the proof of Theorem 6.5.3.

A corollary of Theorem 6.5.1 and Theorem 6.5.3 is the following sufficient condition for oscillation of the solutions of problem (6.5.1), (6.5.2).

Theorem 6.5.4

Let conditions (H6.5) hold in addition to the following conditions

$$\liminf_{t\to\infty} \int_{t_0}^{t} \left(1 - \frac{s}{t}\right)(G(s) + \mu(s)G(s - \sigma) + F(s))\, ds = -\infty \qquad (6.5.26)$$

$$\limsup_{t\to\infty} \int_{t_0}^{t} \left(1 - \frac{s}{t}\right)(G(s) + \mu(s)G(s - \sigma) + F(s))\, ds = +\infty \qquad (6.5.27)$$

for any sufficiently large number t_0, where the functions $G(t)$ and $F(t)$ are defined by (6.5.4). Then each solution $u(x,t)$ of problem (6.5.1), (6.5.2) oscillates in the domain G.

A corollary of Theorem 6.5.2 and Theorem 6.5.3 is the following sufficient condition for oscillation of the solutions of problem (6.5.1), (6.5.3).

Theorem 6.5.5

Let conditions H6.5.1–H6.5.5 hold in addition to the following conditions

$$\mu(t) \geqq 0 \quad for \quad t \geqq 0 \qquad (6.5.28)$$

$$\liminf_{t\to\infty} \int_{t_0}^{t} \left(1 - \frac{s}{t}\right) F_1(s)\, ds = -\infty \qquad (6.5.29)$$

$$\limsup_{t\to\infty} \int_{t_0}^{t} \left(1 - \frac{s}{t}\right) F_1(s)\, ds = +\infty \qquad (6.5.30)$$

for any sufficiently large number t_0, where the function $F_1(t)$ is defined by (6.5.19). Then each solution $u(x,t)$ of problem (6.5.1), (6.5.3) oscillates in the domain G.

Example 6.5.1 Consider the equation

$$u_{tt} + u_{tt}(x, t - \pi) - u_{xx} + u = 2\,e^t(\sin t + \cos t - e^{-\pi}\cos t)\cos x,$$

$$(x, t) \in (0, \pi/2) \times (0, \infty) \equiv G \tag{6.5.31}$$

with boundary solutions

$$u_x(0, t) = 0, \quad u_x(\pi/2, t) = -e^t \sin t, \quad t \geq 0. \tag{6.5.32}$$

A straightforward verification shows that the functions

$$c(x, t, u) = u, \quad f(x, t) = 2\,e^t \cos x(\sin t + \cos t - e^{-\pi}\cos t), \quad g(0, t) = 0,$$

$$g(\pi/2, t) = -e^t \sin t, \quad \lambda(t) \equiv 1, \quad \mu(t) \equiv 0, \quad \gamma(x, t) \equiv 0$$

satisfy conditions (H6.5). Moreover, from (6.5.4) we obtain that

$$G(t) = -\frac{2}{\pi} e^t \sin t, \quad t \geq 0$$

$$F(t) = \frac{4}{\pi} e^t(\sin t + \cos t - e^{-\pi}\cos t), \quad t \geq 0.$$

By straightforward calculations we find that

$$I(t) = \int_{t_0}^{t} \left(1 - \frac{s}{t}\right)(G(s) + \mu(s)G(s - \sigma) + F(s))\,ds$$

$$= \frac{e^t}{t\pi}(2\sin t - 2\,e^{-\pi}\sin t - \cos t) + C$$

where C does not depend on t. Hence

$$\liminf_{t \to \infty} I(t) = -\infty, \quad \limsup_{t \to \infty} I(t) = +\infty$$

that is, conditions (6.5.26), (6.5.27) of Theorem 6.5.4 also hold. Then from Theorem 6.5.4 it follows that each solution of problem (6.5.31), (6.5.32) oscillates in the domain $G = (0, \pi/2) \times (0, \infty)$. For instance, the function $u(x, t) = e^t \sin t \cos x$ is such a solution.

Example 6.5.2   Consider the equation

$$u_{tt} + u_{tt}(x, t - \pi) - [u_{xx} + u_{xx}(x, t - \pi)] + u = f(x, t)$$

$$(x, t) \in (0, \pi) \times (0, \infty) \equiv G,$$

(6.5.33)

where $f(x, t) = e^t \sin x(2 e^{-\pi} \sin t - 2 \sin t + e^{-\pi} \cos t)$

with boundary conditions

$$u(0, t) = u(\pi, t) = 0, \quad t \geq 0.$$

(6.5.34)

A straightforward verification shows that the functions

$$c(x, t, u) = u, \quad f(x, t), \quad \lambda(t) = \mu(t) \equiv 1$$

satisfy conditions H6.5.1–H6.5.5. Moreover, the least eigenvalue of Sturm–Liouville's problem

$$U'' + \alpha U = 0, \quad U(0) = U(\pi) = 0$$

is $\alpha_0 = 1$ and the corresponding eigenfunction is $\varphi(x) = \sqrt{(2/\pi)} \sin x > 0$, $x \in (0, \pi)$. Then from (6.5.19) we find that

$$F_1(t) = \int_0^\pi f(x, t) \sqrt{\frac{2}{\pi}} \sin x \, dx \left(\int_0^\pi \sqrt{\frac{2}{\pi}} \sin x \, dx \right)^{-1}$$

$$= \frac{\pi}{4} e^t (2 e^{-\pi} \sin t - 2 \sin t + e^{-\pi} \cos t).$$

By straightforward calculations we obtain that

$$I_1(t) = \int_{t_0}^t \left(1 - \frac{s}{t}\right) F_1(s) \, ds = \frac{\pi}{4} \frac{e^t}{t} (\cos t - e^{-\pi} \cos t + \tfrac{1}{2} e^{-\pi} \sin t) + C$$

where C does not depend on t. Hence $\lim_{t \to \infty} \inf I_1(t) = -\infty$ and $\lim_{t \to \infty} \sup I_1(t) = \infty$, that is, conditions (6.5.29), (6.5.30) of Theorem 6.5.6 also hold. Then from Theorem 6.5.5 it follows that each solution of problem (6.5.33), (6.5.34) oscillates in the domain $G = (0, \pi) \times (0, \infty)$. For instance, the function $u(x, t) = e^t \sin x \cos t$ is such a solution.

In the subsequent theorems we shall consider some particular cases of equation (6.5.1) for which new sufficient conditions for oscillation of the solutions are obtained.

Suppose that $\lambda(t) \equiv 0$. We shall use the following result of Kusano and Naito [96] concerning differential inequalities of the form

$$(q(t)(p(t)x)')' + h(t, x) \leq r(t), \quad t \geq t_0 \tag{6.5.35}$$

Assume the following conditions fulfilled:
H6.5.13 $p(t), q(t) \in C([t_0, \infty); (0, \infty))$,

$$\int\limits_{t_0}^{\infty} \frac{1}{q(t)} \, dt = \infty$$

H6.5.14 $h(t, x) \in C([t_0, \infty) \times (0, \infty); (0, \infty))$,
$h(t, x)$ is a monotone increasing function of its second argument x.
H6.5.15 $r(t) \in C([t_0, \infty); \mathbb{R})$.

Theorem 6.5.6

Let conditions H6.5.13–H6.5.15 hold and let the differential inequality

$$(q(t)(p(t)x)')' + h(t, x) \leq 0 \tag{6.5.36}$$

have no eventually positive solutions. Moreover, let a function

$$\theta(t) \in C^2([t_0, \infty); \mathbb{R})$$

exist with the following properties:

$\theta(t)$ takes both positive and negative values for arbitrarily large values of t,

$$\tag{6.5.37}$$

$$(q(t)(p(t)\theta(t))')' = r(t), \quad t \geq t_0 \tag{6.5.38}$$

$$\liminf_{t \to \infty} [p(t)\theta(t)] = 0. \tag{6.5.39}$$

Then the differential inequality (6.5.35) has no eventually positive solutions.

A corollary of Theorem 6.5.1 and Theorem 6.5.2 is the following sufficient condition for oscillation of the solutions of problem (6.5.1), (6.5.2) in the case when $\lambda(t) \equiv 0$.

Theorem 6.5.7

Assume the following conditions fulfilled:
   (1) Conditions (H6.5) hold.
   (2) The function $h(u)$ is monotone increasing in the interval $(0, \infty)$.
   (3) The differential inequality

$$x''(t) + p(t)h(x(t)) \leqq 0, \quad t \geqq t_0$$

has no eventually positive solutions.
   (4) There exists a function $\theta(t) \in C^2([t_0, \infty); \mathbb{R})$ with the following properties:
   (a) $\theta(t)$ takes both positive and negative values for arbitrarily large values of t.
   (b) $[\theta(t)]'' = G(t) + \mu(t)G(t - \sigma) + F(t), \quad t \geqq t_1$.
   (c) $\lim_{t \to \infty} \theta(t) = 0$.
   Then each solution $u(x, t)$ of problem (6.5.1), (6.5.2) oscillates in the domain G.

Example 6.5.3   Consider the equation

$$u_{tt}(x, t) - u_{xx}(x, t) - u_{xx}(x, t - \pi) + 2u = f(x, t),$$

$$(x, t) \in (0, \pi/2) \times (0, \infty) \tag{6.5.40}$$

where $f(x, t) = e^{-t} \cos x (3 \sin t - 2 \cos t - e^{\pi} \sin t)$ with boundary condition

$$u_x(0, t) = 0, \quad u_x(\pi/2, t) = -e^{-t} \sin t, \quad t \geqq 0. \tag{6.5.41}$$

A straightforward verification shows that the functions

$$c(x, t, u) = 2u, \quad f(x, t), \quad \mu(t) \equiv 1, \gamma(x, t) \equiv 0, \quad g(0, t) \equiv 0,$$

$$g(\pi/2, t) = -e^{-t} \sin t$$

satisfy conditions (H6.5). Moreover, from (6.5.4) we obtain

$$G(t) = -\frac{2}{\pi} e^{-t} \sin t, \quad t \geqq 0$$

$$F(t) = \frac{2}{\pi} e^{-t} (3 \sin t - 2 \cos t - e^{\pi} \sin t), \quad t \geqq 0.$$

Then

$$I_2(t) = \int_{t_0}^{t} \left(1 - \frac{s}{t}\right)(G(s) + \mu(s)G(s - \sigma) + F(s)) \, ds$$

$$= \int_{t_0}^{t} \left(1 - \frac{s}{t}\right)\frac{2}{\pi} e^{-s}(3 \sin s - 2 \cos s - e^{\pi} \sin s) \, ds$$

from which we obtain immediately that $\lim_{t \to \infty} I_2(t) < \infty$. Hence conditions (6.5.26) and (6.5.27) of Theorem 6.5.4 do not hold. It is easy to check that the differential inequality $x'' + 2x \leq 0$ has no eventually positive solutions. Let

$$\theta(t) = \frac{e^{-t}}{\pi}(2 \sin t + (3 - e^{\pi}) \cos t).$$

Then for $n \in \mathbb{Z}$ we obtain

$$\theta\left(\frac{\pi}{2} + 2n\pi\right) > 0, \quad \theta\left(\frac{3\pi}{2} + 2n\pi\right) < 0.$$

Moreover,

$$[\theta(t)]'' = \frac{2}{\pi} e^{-t}(3 \sin t - 2 \cos t - e^{\pi} \sin t)$$

and $\lim_{t \to \infty} \theta(t) = 0$. Hence the function $\theta(t)$ satisfies condition (4) of Theorem 6.5.7. Then by Theorem 6.5.7 each solution $u(x, t)$ of problem (6.5.40), (6.5.41) oscillates in the domain $G = (0, \pi/2) \times (0, \infty)$. For instance, the function $u(x, t) = e^{-t} \sin t \cos x$ is such a solution.

We shall note that a result analogous to the one of Theorem 6.5.7 can be obtained for problem (6.5.1), (6.5.3).

Suppose that $f(x, t) \equiv 0$, $g(x, t) \equiv 0$. In this case the finding of sufficient conditions for oscillation of the solutions of equation (6.5.1) in the domain G is reduced to the investigation of the oscillatory properties of neutral differential inequalities of the form

$$\frac{d^2}{dt^2}[x(t) + \lambda(t)x(t - \tau)] + q_0(t)x(t) + q(t)x(t - \sigma) + p(t)h(x(t)) \leq 0,$$

$$t \geq t_0. \quad (6.5.42)$$

$$\frac{d^2}{dt^2}[x(t) + \lambda(t)x(t-\tau)] + q_0(t)x(t) + q(t)x(t-\sigma) + p(t)h(x(t)) \geq 0,$$

$$t \geq t_0. \quad (6.5.43)$$

Together with (6.5.42) and (6.5.43) we shall consider the neutral nonlinear differential equation

$$\frac{d^2}{dt^2}[x(t) + \lambda(t)x(t-\tau)] + q_0(t)x(t) + q(t)x(t-\sigma) + p(t)h(x(t)) = 0,$$

$$t \geq t_0. \quad (6.5.44)$$

Assume the following conditions fulfilled:
H6.5.16 $\lambda(t) \in C^2([t_0, \infty); \mathbb{R})$,
         $0 < \lambda_1 \leq \lambda(t) \leq \lambda_2$ for $t \geq t_0$, $\lambda_1, \lambda_2 = \text{const.}$
H6.5.17 $q_0(t), q(t) \in C([t_0, \infty); [0, \infty))$.
H6.5.18 $p(t) \in C([t_0, \infty); (0, \infty))$.
H6.5.19 $h(u) \in C(\mathbb{R}; \mathbb{R})$, $h(-u) = -h(u)$.
$h(u)$ is a positive and monotone increasing function in the interval $(0, \infty)$.

Theorem 6.5.8

Let the following conditions hold:
  (1) Conditions H6.5.16–H6.5.19 are satisfied.
  (2) For any closed and measurable set $E \subset [t_0, \infty)$ for which meas $(E \cap [t, t+2\tau]) \geq \tau, t \in [t_0, \infty)$ the following condition holds

$$\int_E p(t)\,dt = \infty. \quad (6.5.45)$$

Then:
  (i)   the differential inequality (6.5.42) has no eventually positive solutions;
  (ii)   the differential inequality (6.5.43) has no eventually negative solutions;
  (iii)   all solutions of the differential equation (6.5.44) oscillate.

Proof

(i) Let $x(t)$ be an eventually positive solution of the differential inequality (6.5.42). Then there exists a number $t_1 \geq t_0$ such that $x(t) > 0$, $x(t-\tau) > 0$

and $x(t - \sigma) > 0$ for $t \geq t_1$. From conditions H6.5.17—H6.5.19 and (6.5.42) it follows that

$$\frac{d^2}{dt^2}[x(t) + \lambda(t)x(t - \tau)] \leq -q_0(t)x(t) - q(t)x(t - \sigma) - p(t)h(x(t))$$

$$\leq -p(t)h(x(t)) < 0, \quad t \geq t_1. \qquad (6.5.46)$$

Hence the function $d/dt[x(t) + \lambda(t)x(t - \tau)]$ is monotone decreasing in the interval $[t_1, \infty)$. Suppose that there exists a number $t_2 \geq t_1$ such that $d/dt[x(t_2) + \lambda(t_2)x(t_2 - \tau)] = -C < 0$. Then for any point $t \geq t_2$ the following inequality holds

$$\frac{d}{dt}[x(t) + \lambda(t)x(t - \tau)] \leq \frac{d}{dt}[x(t_2) + \lambda(t_2)x(t_2 - \tau)] = -C.$$

Integrate the last inequality over the interval $[t_2, t]$, $t > t_2$ and obtain

$$x(t) + \lambda(t)x(t - \tau) \leq x(t_2) + \lambda(t_2)x(t_2 - \tau) - C(t - t_2).$$

Hence $\lim_{t \to \infty} \sup[x(t) + \lambda(t)x(t - \tau)] \leq 0$ which contradicts the assumption that $x(t)$ is an eventually positive solution. Hence

$$\frac{d}{dt}[x(t) + \lambda(t)x(t - \tau)] \geq 0, \quad t \geq t_1 \qquad (6.5.47)$$

from which we obtain that $x(t) + \lambda(t)x(t - \tau) \geq C_1 > 0$ for $t \geq t_1$. From Lemma 6.2.3 it follows that there exists a closed and measurable set $E \subset [t_1, \infty)$ and a constant $C_2 > 0$ such that $x(t) \geq C_2$ for $t \in E$ and $\text{meas}(E \cap [t, t + 2\tau]) \geq \tau$ for $t \geq t_1$. Then from condition H6.5.19 it follows that

$$h(x(t)) \geq h(C_2) = C_3 > 0 \quad \text{for} \quad t \in E.$$

Integrate both sides of inequality (6.5.46) over the interval $[t_1, t]$, $t \geq t_1$ and using (6.5.47) obtain

$$C_3 \int_{E \cap [t_1, t]} p(s)\, ds \leq \int_{t_1}^{t} p(s)h(x(s))\, ds \leq \frac{d}{dt}[x(t_1) + \lambda(t_1)x(t_1 - \tau)]$$

$$-\frac{d}{dt}[x(t) + \lambda(t)x(t - \tau)]$$

$$\leq \frac{d}{dt}[x(t_1) + \lambda(t_1)x(t_1 - \tau)] = C_4.$$

For $t \to \infty$ from the above inequality it follows that $\int_E p(t)\,dt < \infty$ which contradicts condition (6.5.45). Thus assertion (i) of Theorem 6.5.8 is proved.

(ii) The proof follows immediately from the fact that if $x(t)$ is an eventually negative solution of the differential inequality (6.5.43), then $-x(t)$ is an eventually positive solution of the differential inequality (6.5.42).

(iii) The proof follows immediately from assertions (i) and (ii).

A corollary of Theorem 6.5.1, Theorem 6.5.2 and Theorem 6.5.8 is the following sufficient condition for oscillation of the solutions of problem (6.5.1), (6.5.2) or (6.5.1), (6.5.3) in the case when $f(x,t) \equiv 0$ and $g(x,t) \equiv 0$.

Theorem 6.5.9

Let the following conditions be satisfied:
(1) Conditions ($H6.5$) hold.
(2) $0 < \lambda_1 \leq \lambda(t) \leq \lambda_2$, $t \geq t_0$, $\lambda_1, \lambda_2 = $ const.
(3) $h(-u) = -h(u)$, $u \in \mathbb{R}$; $h(u)$ is a monotone increasing function in the interval $(0, \infty)$.
(4) For any closed and measurable set $E \subset [t_0, \infty)$ for which meas *$(E \cap [t, t + 2\tau] \geq \tau$, $t \in [t_0, \infty)$ the following condition holds*

$$\int_E p(t)\,dt = \infty.$$

Then each solution $u(x,t)$ of problem (6.5.1), (6.5.2) or (6.5.1), (6.5.3) oscillates in the domain G.

Suppose that $\lambda(t) \equiv 0$, $g(x,t) \equiv 0$, $\mu(t) = \mu = $ const. > 0. In this case results are obtained on the distribution of the zeros of the solutions of nonlinear hyperbolic equations of neutral type of the form

$$u_{tt}(x,t) - [\Delta u(x,t) + \mu\,\Delta u(x, t - \sigma)] + c(x,t,u) = f(x,t),$$

$$(6.5.49)$$

$$(x,t) \in \Omega \times (0, \infty) \equiv G$$

with boundary condition

$$\frac{\partial u}{\partial n} + \gamma(x,t)u = 0, \quad (x,t) \in \partial\Omega \times [0, \infty) \qquad (6.5.50)$$

or

$$u = 0, \quad (x,t) \in \partial\Omega \times [0, \infty). \qquad (6.5.51)$$

Assume the following conditions fulfilled:

H6.5.20 $c(x, t, \eta) \in C(\bar{G} \times \mathbb{R}; \mathbb{R})$.

H6.5.21 $c(x, t, \eta) \geq k^2 \eta$, $(x, t, \eta) \in G \times [0, \infty)$, where $k = $ const. > 0.

H6.5.22 $c(x, t, \eta) \leq k^2 \eta$, $(x, t, \eta) \in G \times (-\infty, 0)$.

H6.5.23 $f(x, t) \in C(\bar{G}; \mathbb{R})$.

H6.5.24 $\gamma(x, t) \in C(\partial\Omega \times [0, \infty); [0, \infty))$.

In the proof of the subsequent two theorems we shall use the following result concerning differential inequalities of the form

$$x''(t) + \omega^2 x(t) \leq r(t), \quad \omega = \text{const.} > 0. \tag{6.5.52}$$

Lemma 6.5.3

Let a number s exist such that the following inequality holds

$$\int_{s}^{s + \pi/\omega} \sin \omega(t - s) r(t) \, dt \leq 0. \tag{6.5.53}$$

Then the differential inequality (6.5.52) has no solutions which are positive in the interval $[s, s + \pi/\omega)$.

Proof

Suppose that this is not true and let $x(t)$ be a solution of the differential inequality (6.5.52) which is positive in the interval $[s, s + \pi/\omega)$. Multiply both sides of inequality (6.5.52) by the function $\sin \omega(t - s)$, integrate over the interval $[s, s + \pi/\omega]$ and obtain

$$\omega \left[x(s) + x\left(s + \frac{\pi}{\omega}\right) \right] \leq \int_{s}^{s + \pi/\omega} \sin \omega(t - s) r(t) \, dt.$$

From the assumption it follows that the left-hand side of the last inequality is positive. Hence

$$\int_{s}^{s + \pi/\omega} \sin \omega(t - s) r(t) \, dt > 0$$

which contradicts condition (6.5.53). This completes the proof of Lemma 6.5.3.

Theorem 6.5.10

Let conditions H6.5.20–H6.5.24 hold and let a number $s \geq \sigma$ exist such that

$$H_1(t) \equiv \int_s^{s+\pi/k} F(t) \sin k(t-s)\, dt = 0 \qquad (6.5.54)$$

where the function $F(t)$ is defined by (6.5.4).
  Then each solution $u(x, t)$ of problem (6.5.49), (6.5.50) has a zero in the domain $\Omega \times (s - \sigma, s + \pi/k)$.

Proof

Suppose that this is not true and let $u(x, t)$ be a solution of problem (6.5.49), (6.5.50) without a zero in $\Omega \times (s - \sigma, s + \pi/k)$. Suppose that

$$u(x, t) > 0 \quad \text{for} \quad (x, t) \in \Omega \times (s - \sigma, s + \pi/k) \qquad (6.5.55)$$

Integrate both sides of equation (6.5.49) with respect to x over the domain Ω and obtain

$$\frac{d^2}{dt^2} \int_\Omega u(x, t)\, dx - \left[\int_\Omega \Delta u(x, t)\, dx + \mu \int_\Omega \Delta u(x, t - \sigma)\, dx \right]$$

$$+ \int_\Omega c(x, t, u(x, t))\, dx = \int_\Omega f(x, t)\, dx. \qquad (6.5.56)$$

From Green's formula, condition H6.5.24 and (H6.5.55), it follows that

$$\int_\Omega \Delta u(x, t)\, dx = \int_{\partial\Omega} \frac{\partial u}{\partial n}\, ds = - \int_{\partial\Omega} \gamma(x, t) u(x, t)\, ds \leq 0, \quad t \in \left[s, s + \frac{\pi}{k} \right)$$

$$(6.5.57)$$

$$\int_\Omega \Delta u(x, t - \sigma)\, dx = \int_{\partial\Omega} \frac{\partial u}{\partial n}(x, t - \sigma)\, ds = - \int_{\partial\Omega} \gamma(x, t - \sigma) u(x, t - \sigma)\, ds \leq 0,$$

$$t \in \left[s, s + \frac{\pi}{k} \right). \qquad (6.5.58)$$

Moreover, from condition H6.5.21 and (6.5.55) it follows that

$$c(x, t, u(x, t)) \geq k^2 u(x, t), \quad t \in \left[s, s + \frac{\pi}{k} \right). \qquad (6.5.59)$$

Using (6.5.57)–(6.5.59), from (6.5.56) we obtain

$$v''(t) + k^2 v(t) \leq F(t), \quad t \in \left[s, s + \frac{\pi}{k} \right) \qquad (6.5.60)$$

where the function $v(t)$ is defined by (6.5.5). From Lemma 6.5.3 and condition (6.5.54) it follows that the differential inequality (6.5.60) has no solutions which are positive in the interval $[s, s + \pi/k)$. On the other hand, from (6.5.55) it follows that the function $v(t)$ is a solution of (6.5.60) which is positive in $[s, s + \pi/k)$ in contradiction with what was said above. The case when $u(x, t) < 0$ in $\Omega \times (s - \sigma, s + \pi/k)$ is considered analogously.

This completes the proof of Theorem 6.5.10.

Definition 6.5.2   The function $H_1(t) \in C(\mathbb{R}; \mathbb{R})$ is said to oscillate if there exists a sequence of zeros $\{t_n\}_{n=1}^{\infty}$ of $H_1(t)$ satisfying the equality $\lim_{n \to \infty} t_n = +\infty$.

Corollary 6.5.1   *Let conditions H6.5.20–H6.5.24 hold and let the function $H_1(t)$ oscillate. Then each solution $u(x, t)$ of problem (6.5.49), (6.5.50) oscillates in the domain G.*

Example 6.5.4   Consider the equation

$$u_{tt}(x, t) - [u_{xx}(x, t) + u_{xx}(x, t - \pi)] + 4u(x, t) = 3 \sin t \sin \left(\frac{x}{2} + \frac{\pi}{4} \right),$$

$$(x, t) \in (0, \pi) \times (0, \infty) \quad (6.5.61)$$

with boundary condition

$$-u_x(0, t) + \tfrac{1}{2} u(0, t) = u_x(\pi, t) + \tfrac{1}{2} u(\pi, t) = 0, \quad t \geq 0. \qquad (6.5.62)$$

A straightforward verification shows that the functions

$$c(x, t, u) = 4u, \quad f(x, t) = 3 \sin t \sin \left(\frac{x}{2} + \frac{\pi}{4} \right), \quad \mu = 1$$

satisfy conditions H6.5.20–H6.5.24. By straightforward calculations we obtain that

$$F(t) = 6\sqrt{2}\sin t$$

$$H_1(s) = \int_s^{s+\pi/2} 6\sqrt{2}\sin t \sin 2(t-s)\,dt = 8\sin\left(s+\frac{\pi}{4}\right).$$

Then from $H_1(s) = 0$ it follows that $s = s_n = \frac{3}{4}\pi + n\pi > \pi$, $n = 1,2,3,\ldots$. Hence condition (6.5.54) of Theorem 6.5.10 is fulfilled. Then by this theorem each solution $u(x,t)$ of problem (6.5.61), (6.5.62) has a zero in the domain $\Omega \times (s_n - \pi, s_n + \pi/2)$. For instance, the function $u(x,t) = \sin t \sin[(x/2) + (\pi/4)]$ is such a solution.

Theorem 6.5.11

Let conditions H6.5.20–H6.5.24 hold and let a number $s \geq \sigma$ exist such that

$$H_2(t) \equiv \int_s^{s+\pi/\omega} F_1(t)\sin\omega(t-s)\,dt = 0 \qquad (6.5.63)$$

where $\omega = \sqrt{(\alpha_0 + k^2)}$ and the function $F_1(t)$ is defined by (6.5.19).
Then each solution $u(x,t)$ of problem (6.5.49), (6.5.51) has a zero in the domain $\Omega \times (s - \sigma, s + \pi/\omega)$.

Proof

Suppose that this is not true and let $u(x,t)$ be a solution of problem (6.5.49), (6.5.51) without a zero in the domain $\Omega \times (s - \sigma, s + \pi/\omega)$. Suppose that

$$u(x,t) > 0 \quad \text{for} \quad (x,t) \in \Omega \times (s - \sigma, s + \pi/\omega). \qquad (6.5.64)$$

Multiply both sides of equation (6.5.49) by the eigenfunction $\varphi(x)$ of problem (6.1.12), (6.1.13), integrate with respect to x over the domain Ω and obtain

$$\frac{d^2}{dt^2}\int_\Omega u(x,t)\varphi(x)\,dx - \left[\int_\Omega \Delta u(x,t)\varphi(x)\,dx + \mu\int_\Omega \Delta u(x,t-\sigma)\varphi(x)\,dx\right]$$

$$+ \int_\Omega c(x,t,u(x,t))\varphi(x)\,dx = \int_\Omega f(x,t)\varphi(x)\,dx. \qquad (6.5.65)$$

From Green's formula it follows that

$$\int_\Omega \Delta u(x, t)\varphi(x)\, dx = \int_\Omega u(x, t)\, \Delta\varphi(x)\, dx = -\alpha_0 \int_\Omega u(x, t)\varphi(x)\, dx$$

$$(6.5.66)$$

$$\int_\Omega \Delta u(x, t - \sigma)\varphi(x)\, dx = \int_\Omega u(x, t - \sigma)\, \Delta\varphi(x)\, dx$$

$$= -\alpha_0 \int_\Omega u(x, t - \sigma)\varphi(x)\, dx \qquad (6.5.67)$$

where α_0 is the least eigenvalue of problem (6.1.12), (6.1.13). Moreover, from condition H6.5.21 and (6.5.64) it follows that

$$\int_\Omega c(x, t, u(x, t))\varphi(x)\, dx \geq k^2 \int_\Omega u(x, t)\varphi(x)\, dx, \quad t \in \left[s, s + \frac{\pi}{\omega} \right).$$

$$(6.5.68)$$

Using (6.5.66)–(6.5.68), from (6.5.65) we obtain

$$w''(t) + (\alpha_0 + k^2)w(t) + \alpha_0\mu w(t - \sigma) \leq F_1(t), \quad t \in \left[s, s + \frac{\pi}{\omega} \right)$$

where the function $w(t)$ is defined by (6.5.13). Since from (6.5.64) it follows that $w(t - \sigma) \geq 0$, then from the last inequality we obtain

$$w''(t) + \omega^2 w(t) \leq F_1(t), \quad t \in \left[s, s + \frac{\pi}{\omega} \right). \qquad (6.5.69)$$

From Lemma 6.5.3 and condition 6.5.63 it follows that the differential inequality (6.5.69) has no solutions which are positive in the interval $[s, s + \pi/\omega)$. On the other hand, from (6.5.64) it follows that the function $w(t)$ is a solution of (6.5.69) which is positive in $[s, s + \pi/\omega)$ in contradiction with what was said above. The case when $u(x, t) < 0$ in $\Omega \times (s - \sigma, s + \pi/\omega)$ is considered analogously.

This completes the proof of Theorem 6.5.11.

Corollary 6.5.2 *Let conditions H6.5.20–H6.5.24 hold and let the function $H_2(t)$ oscillate. Then each solution $u(x, t)$ of problem (6.5.49), (6.5.51) oscillates in the domain G.*

Example 6.5.5 Consider the equation

$$u_{tt}(x, t) - [u_{xx}(x, t) + u_{xx}(x, t - \pi/2)] + u(x, t) = (\sin t - \cos t) \sin x,$$

$$(x, t) \in (0, \pi) \times (0, \infty) \quad (6.5.70)$$

with boundary condition

$$u(0, t) = u(\pi, t) = 0, \quad t \geq 0. \quad (6.5.71)$$

A straightforward verification shows that the functions

$$c(x, t, u) = u, \quad f(x, t) = (\sin t - \cos t) \sin x, \quad \mu = 1$$

satisfy conditions H6.5.20–H6.5.24. By straightforward calculations we obtain that

$$F_1(t) = \frac{\pi}{2} (\sin t - \cos t)$$

$$H_2(s) = \int_s^{s + \pi/(\sqrt{2})} F_1(t) \sin \sqrt{2}(t - s) \, dt = 2\pi \sin \left(s - \frac{\pi}{4} \right) + \frac{\pi}{2\sqrt{2}} \cos \frac{\pi}{2\sqrt{2}}.$$

Then from $H_2(s) = 0$ it follows that $s = s_n = (\frac{5}{4} - 1/\sqrt{2})\pi + n\pi > \pi/2$, $n = 0, 1, 2, \ldots$. Hence condition (6.5.63) of Theorem 6.5.11 is satisfied. Then in virtue of this theorem each solution $u(x, t)$ of problem (6.5.70), (6.5.71) has a zero in the domain $\Omega \times (s_n - \pi/2, s_n + \pi/\sqrt{2})$. For instance, the function $u(x, t) = \sin t \sin x$ is such a solution.

6.6 Hyperbolic differential equations with 'maxima'

In this section sufficient conditions are obtained for oscillation of the solutions of neutral hyperbolic differential equations with 'maxima' of the form

$$\frac{\partial^2}{\partial t^2} [u(x, t) + \lambda(t) u(x, \tau(t))] - [\Delta u(x, t) + \mu(t) \Delta u(x, \sigma(t))]$$

$$+ c \left(x, t, u(x, t), \max_{s \in M(t)} u(x, s) \right) = f(x, t),$$

$$(x, t) \in \Omega \times (0, \infty) \equiv G \quad (6.6.1)$$

where $\Delta u(x, t) = \sum_{i=1}^{n} u_{x_i x_i}(x, t)$, $M(t) \subset [0, \infty)$ for $t \geq t_0$ and Ω is a bounded domain in $\mathbb{R}^n$ with a piecewise smooth boundary. Consider boundary conditions of the form

$$u(x, t) = g(x, t), \quad (x, t) \in \partial\Omega \times [0, \infty) \tag{6.6.2}$$

$$\frac{\partial u}{\partial n} + \gamma(x, t)u = h(x, t), \quad (x, t) \in \partial\Omega \times [0, \infty). \tag{6.6.3}$$

We shall say that conditions (H6.6) are met if the following conditions hold:

H6.6.1 $\lambda(t) \in C^2([0, \infty); [0, \infty))$.

H6.6.2 $\mu(t) \in C([0, \infty); [0, \infty))$.

H6.6.3 $\tau(t), \sigma(t) \in C^2([0, \infty); \mathbb{R})$,

$$\lim_{t \to \infty} \tau(t) = \lim_{t \to \infty} \sigma(t) = \infty.$$

H6.6.4 $c(x, t, \beta, \eta) \in C(G \times \mathbb{R}^2; \mathbb{R})$,
$c(x, t, \beta, \eta) \geq 0$ for $(x, t) \in G$, $\beta \geq 0$, $\eta \geq 0$,
$c(x, t, \beta, \eta) \leq 0$ for $(x, t) \in G$, $\beta \leq 0$, $\eta \leq 0$,

H6.6.5 $f(x, t) \in C(G; \mathbb{R})$.

H6.6.6 $g(x, t) \in C(\partial\Omega \times [0, \infty); \mathbb{R})$.

H6.6.7 $M(t)$ is a closed and bounded set for any $t \in [0, \infty)$ and $\lim_{t \to \infty} v(t) = \infty$, where $v(t) = \min_{s \in M(t)} s$.

H6.6.8 $h(x, t) \in C(\partial\Omega \times [0, \infty); \mathbb{R})$,

$$\gamma(x, t) \in C(\partial\Omega \times [0, \infty); [0, \infty)).$$

Definition 6.6.1 The solution $u(x, t) \in C^2(G) \cap C^1(\bar{G})$ of problem (6.6.1), (6.6.2) ((6.6.1), (6.6.3)) is said to oscillate in the domain G if for any positive number ρ there exists a point $(x', t') \in \Omega \times [\rho, \infty)$ such that the equality $u(x', t') = 0$ holds.

We shall note that under the absence of 'maxima' in the work of Kreith *et al* [91] conditions were obtained for oscillation of the solutions of problem (6.6.1), (6.6.3) in the case when $\lambda(t) \equiv 0$, $\mu(t) \equiv 0$ and $\gamma(x, t) \equiv 0$. Further generalizations were obtained by Yoshida [155].

Let α_0 be the least eigenvalue and let $\varphi(x)$ be the corresponding eigenfunction of problem (6.1.12), (6.1.13). It is well known [150] that $\alpha_0 > 0$ and $\varphi(x)$ can be chosen so that $\varphi(x) > 0$ for $x \in \Omega$.

With each solution $u(x, t) \in C^2(G) \cap C^1(\bar{G})$ of problem (6.6.1), (6.6.2) we associate the function

$$w(t) = \int_{\Omega} u(x, t)\varphi(x)\,dx \left(\int_{\Omega} \varphi(x)\,dx \right)^{-1}, \quad t \geq 0. \tag{6.6.4}$$

Lemma 6.6.1

Let conditions $(H6.6)$ hold and let $u(x, t)$ be a positive solution of problem (6.6.1), (6.6.2) in the domain G. Then the function $w(t)$ defined by (6.6.4) satisfies the differential inequality

$$\frac{d^2}{dt^2}[w(t) + \lambda(t)w(\tau(t))] + \alpha_0[w(t) + \mu(t)w(\sigma(t))] \leq \Phi(t), \quad t \geq t_1$$

$$\tag{6.6.5}$$

where

$$\Phi(t) = \left[\int_{\Omega} f(x, t)\varphi(x)\,dx - \int_{\partial\Omega} (g(x, t) + \mu(t)g(x, \sigma(t))) \frac{\partial\varphi}{\partial n}\,ds \right]$$

$$\times \left(\int_{\Omega} \varphi(x)\,dx \right)^{-1} \tag{6.6.6}$$

Proof

Let $u(x, t)$ be a solution of problem (6.6.1), (6.6.2) for which $u(x, t) > 0$ for $(x, t) \in G$. Then from conditions H6.3.3 and H6.6.7 it follows that there exists a number $t_1 \geq 0$ such that $u(x, \tau(t)) > 0$, $u(x, \sigma(t)) > 0$ and $\max_{s \in M(t)} u(x, s) > 0$ for $(x, t) \in \Omega \times [t_1, \infty)$. Multiply both sides of equation (6.6.1) by the function $\varphi(x)$ and integrate with respect to x over the domain

Ω. For $t \geq t_1$ we obtain

$$\frac{d^2}{dt^2} \left[\int_\Omega u(x, t)\varphi(x)\,dx + \lambda(t) \int_\Omega u(x, \tau(t))\varphi(x)\,dx \right]$$

$$- \left[\int_\Omega \Delta u(x, t)\varphi(x)\,dx + \mu(t) \int_\Omega \Delta u(x, \sigma(t))\varphi(x)\,dx \right]$$

$$+ \int_\Omega c\left(x, t, u(x, t), \max_{s \in M(t)} u(x, s) \right) \varphi(x)\,dx = \int_\Omega f(x, t)\varphi(x)\,dx. \qquad (6.6.7)$$

From Green's formula it follows that

$$\int_\Omega \Delta u(x, t)\varphi(x)\,dx = - \int_{\partial\Omega} u(x, t)\frac{\partial\varphi}{\partial n}\,ds + \int_\Omega u(x, t)\,\Delta\varphi(x)\,dx$$

$$= - \int_{\partial\Omega} g(x, t)\frac{\partial\varphi}{\partial n}\,ds - \alpha_0 \int_\Omega u(x, t)\varphi(x)\,dx. \qquad (6.6.8)$$

$$\int_\Omega \Delta u(x, \sigma(t))\varphi(x)\,dx = - \int_{\partial\Omega} u(x, \sigma(t))\frac{\partial\varphi}{\partial n}\,ds + \int_\Omega u(x, \sigma(t))\,\Delta\varphi(x)\,dx$$

$$= - \int_{\partial\Omega} g(x, \sigma(t))\frac{\partial\varphi}{\partial n}\,ds$$

$$- \alpha_0 \int_\Omega u(x, \sigma(t))\varphi(x)\,dx. \qquad (6.6.9)$$

From condition H6.6.4 it follows that

$$\int_\Omega c\left(x, t, u(x, t), \max_{s \in M(t)} u(x, s) \right) \varphi(x)\,dx \geq 0. \qquad (6.6.10)$$

Using (6.6.8)–(6.6.10), from (6.6.7) we obtain

$$\frac{d^2}{dt^2}[w(t) + \lambda(t)w(\tau(t))] + \alpha_0[w(t) + \mu(t)w(\sigma(t))]$$

$$\leqq \left[\int_\Omega f(x,t)\varphi(x)\,dx - \int_{\partial\Omega} g(x,t)\frac{\partial\varphi}{\partial n}\,ds - \mu(t)\int_{\partial\Omega} g(x,\sigma(t))\frac{\partial\varphi}{\partial n}\,ds \right]$$

$$\times \left(\int_\Omega \varphi(x)\,dx \right)^{-1}$$

which was to be proved.

Analogously to the proof of Lemma 6.6.1 the following lemma can be proved.

Lemma 6.6.2

Let conditions (H6.6) hold and let $u(x,t)$ be a negative solution of problem (6.6.1), (6.6.2) in the domain G. Then the function $w_1(t) = -w(t)$ satisfies the differential inequality

$$\frac{d^2}{dt^2}[w_1(t) + \lambda(t)w_1(\tau(t))] + \alpha_0[w_1(t) + \mu(t)w_1(\sigma(t))] \leqq -\Phi(t), \quad t \geqq t_1$$

$$(6.6.11)$$

where the function $\Phi(t)$ is defined by (6.6.6).

Definition 6.2.2 The solution $w(t)\in C^2([t_1,\infty);\mathbb{R})$ of the differential inequality (6.6.5) is said to be eventually positive (negative) if there exists a number $t_2 \geqq t_1$ such that the inequality $w(t) > 0$ $(w(t) < 0)$ holds for all $t \geqq t_2$.

Theorem 6.6.1

Let conditions (H6.6) hold and let the differential inequalities (6.6.5) and (6.6.11) have no eventually positive solutions. Then each solution $u(x,t)$ of problem (6.6.1), (6.6.2) oscillates in the domain G.

Proof

Let $\rho > 0$ be an arbitrary number. Suppose that the assertion is not true and let $u(x, t)$ be a solution of problem (6.6.1), (6.6.2) without a zero in the domain $G_\rho = \Omega \times [\rho, \infty)$. If $u(x, t) > 0$ for $(x, t) \in G_\rho$, then from Lemma 6.6.1 it follows that the function $w(t)$ defined by (6.6.4) is a positive solution of inequality (6.6.5) which contradicts the condition of the theorem. If $u(x, t) < 0$ for $(x, t) \in G_\rho$, from Lemma 6.6.2 it follows that the function $w_1(t) = -w(t)$ is a positive solution of inequality (6.6.11) which also contradicts the condition of the theorem. Thus Theorem 6.6.1 is proved.

Now we shall investigate the oscillatory properties of the solutions of problem (6.6.1), (6.6.3). With each solution $u(x, t) \in C^2(G) \cap C^1(\bar{G})$ of problem (6.6.1), (6.6.3) we associate the function

$$v(t) = \int_\Omega u(x, t) \, dx \left(\int_\Omega dx \right)^{-1}, \quad t \geq 0. \tag{6.6.12}$$

Lemma 6.6.3

Let conditions (H6.6) hold and let $u(x, t)$ be a positive solution of problem (6.6.1), (6.6.3) in the domain G. Then the function $v(t)$ defined by (6.6.12) satisfies the differential inequality

$$\frac{d^2}{dt^2}[v(t) + \lambda(t)v(\tau(t))] \leq \Phi_1(t), \quad t \geq t_1 \tag{6.6.13}$$

where

$$\Phi_1(t) = \left[\int_\Omega f(x, t) \, dx + \int_{\partial\Omega} (h(x, t) + \mu(t)h(x, \sigma(t))) \, ds \right] \left(\int_\Omega dx \right)^{-1}.$$

$$\tag{6.6.14}$$

Proof

Let $u(x, t)$ be a solution of problem (6.6.1), (6.6.3) for which $u(x, t) > 0$ for $(x, t) \in G$. Then from conditions H6.6.3 and H6.6.7 it follows that there exists a number $t_1 \geq 0$ such that $u(x, \tau(t)) > 0$, $u(x, \sigma(t)) > 0$ and

$\max_{s \in M(t)} u(x, s) > 0$ for $(x, t) \in \Omega \times [t_1, \infty)$. Integrate both sides of equation (6.6.1) with respect to x over the domain Ω and for $t \geq t_1$ obtain

$$\frac{d^2}{dt^2} \left[\int_\Omega u(x, t) \, dx + \lambda(t) \int_\Omega u(x, \tau(t)) \, dx \right]$$

$$- \left[\int_\Omega \Delta u(x, t) \, dx + \mu(t) \int_\Omega \Delta u(x, \sigma(t)) \, dx \right]$$

$$+ \int_\Omega c\left(x, t, u(x, t), \max_{s \in M(t)} u(x, s) \right) dx = \int_\Omega f(x, t) \, dx. \qquad (6.6.15)$$

From Green's formula and condition H6.6.8 it follows that

$$\int_\Omega \Delta u(x, t) \, dx = \int_{\partial\Omega} \frac{\partial u}{\partial n} \, ds = \int_{\partial\Omega} [h(x, t) - \gamma(x, t)u] \, ds$$

$$\leq \int_{\partial\Omega} h(x, t) \, ds \qquad (6.6.16)$$

$$\int_\Omega \Delta u(x, \sigma(t)) \, dx = \int_{\partial\Omega} \frac{\partial u}{\partial n}(x, \sigma(t)) \, ds$$

$$= \int_{\partial\Omega} [h(x, \sigma(t)) - \gamma(x, \sigma(t))u(x, \sigma(t))] \, ds$$

$$\leq \int_{\partial\Omega} h(x, \sigma(t)) \, ds. \qquad (6.6.17)$$

From condition H6.6.4 it follows that

$$\int_\Omega c\left(x, t, u(x, t), \max_{s \in M(t)} u(x, s) \right) dx \geq 0. \qquad (6.6.18)$$

Using (6.6.16)–(6.6.18), from (6.6.15) we obtain

$$\frac{d^2}{dt^2}[v(t) + \lambda(t)v(\tau(t))]$$

$$\leq \left[\int_\Omega f(x,t)\,dx + \int_{\partial\Omega} h(x,t)\,ds + \mu(t) \int_{\partial\Omega} h(x,\sigma(t))\,ds \right] \left(\int_\Omega dx \right)^{-1},$$

which was to be proved.

Analogously the following lemma can be proved:

Lemma 6.6.4

Let conditions (H6.6) hold and let $u(x,t)$ be a negative solution of problem (6.6.1), (6.6.3) in the domain G. Then the function $v_1(t) = -v(t)$ satisfies the differential inequality

$$\frac{d^2}{dt^2}[v_1(t) + \lambda(t)v_1(\tau(t))] \leq -\Phi_1(t), \quad t \geq t_1 \qquad (6.6.19)$$

where the function $\Phi_1(t)$ is defined by (6.6.14).

By means of Lemma 6.6.3 and Lemma 6.6.4 the following theorem is proved.

Theorem 6.6.2

Let conditions (H6.6) hold and let the differential inequalities (6.6.13) and (6.6.19) have no eventually positive solutions. Then each solution $u(x,t)$ of problem (6.6.1), (6.6.3) oscillates in the domain G.

From the theorems proved above it follows that the finding of sufficient conditions for oscillation of the solutions of equation (6.6.1) in the domain G is reduced to the investigation of the oscillatory properties of neutral differential inequalities of the form

$$\frac{d^2}{dt^2}[y(t) + \lambda(t)y(\tau(t))] + q_1(t)y(t) + q_2(t)y(\sigma(t)) \leq F(t), \quad t \geq t_0$$

$$(6.6.20)$$

Assume the following conditions fulfilled:

H6.6.9 $\lambda(t) \in C^2([t_0, \infty); [0, \infty))$.

H6.6.10 $\tau(t) \in C^2([t_0, \infty); \mathbb{R})$, $\sigma(t) \in C([t_0, \infty); \mathbb{R})$, $\lim_{t\to\infty} \tau(t) = \lim_{t\to\infty} \sigma(t) = \infty$.

H6.6.11 $q_i(t) \in C([t_0, \infty); [0, \infty)), i = 1, 2.$
H6.6.12 $F(t) \in C([t_0, \infty); \mathbb{R}).$

Theorem 6.6.3

Let conditions H6.6.9–H6.6.12 hold in addition to the condition

$$\liminf_{t \to \infty} \frac{1}{t - t_1} \int_{t_1}^{t} (t - s)F(s) \, ds = -\infty \qquad (6.6.21)$$

for any number $t_1 \geq t_0$. Then the differential inequality (6.6.20) has no eventually positive solutions.

Proof

Suppose that this is not true and let $y(t)$ be an eventually positive solution of (6.6.20). Then from condition H6.6.10 it follows that there exists a number $t_1 \geq t_0$ such that $y(t) > 0$, $y(\tau(t)) > 0$ and $y(\sigma(t)) > 0$ for $t \geq t_1$. Integrate both sides of inequality (6.6.20) twice over the interval $[t_1, t]$, $t > t_1$ and using condition H6.6.11 obtain

$$y(t) + \lambda(t)y(\tau(t)) \leq C_1 + C_2(t - t_1) + \int_{t_1}^{t}\left[\int_{t_1}^{\rho} F(s) \, ds \right] d\rho$$

where $C_1, C_2 = $ const. Since

$$\int_{t_1}^{t}\left[\int_{t_1}^{\rho} F(s) \, ds \right] d\rho = \int_{t_1}^{t} (t - s)F(s) \, ds$$

from the last inequality it follows that

$$\frac{y(t) + \lambda(t)y(\tau(t))}{t - t_1} \leq \frac{C_1}{t - t_1} + C_2 + \frac{1}{t - t_1} \int_{t_1}^{t} (t - s)F(s) \, ds. \qquad (6.6.22)$$

Then for $t \to \infty$ from (6.6.22), using condition (6.6.21), we obtain

$$\liminf_{t \to \infty} \frac{y(t) + \lambda(t)y(\tau(t))}{t - t_1} = -\infty. \qquad (6.6.23)$$

On the other hand, using condition H6.6.9 and $y(t) > 0$, $y(\tau(t)) > 0$ for $t \geq t_1$, we obtain that

$$\lim_{t \to \infty} \inf \frac{y(t) + \lambda(t)y(\tau(t))}{t - t_1} \geq 0$$

which contradicts equality (6.6.23).

This completes the proof of Theorem 6.6.3.

A corollary of Theorem 6.6.1 and Theorem 6.6.3 is the following sufficient condition for oscillation of the solutions of problem (6.6.1), (6.6.2).

Theorem 6.6.4

Let conditions (H6.6) hold in addition to the conditions:

$$\lim_{t \to \infty} \inf \int_{t_0}^{t} \left(1 - \frac{s}{t} \right) \Phi(s)\, ds = -\infty \qquad (6.6.24)$$

$$\lim_{t \to \infty} \sup \int_{t_0}^{t} \left(1 - \frac{s}{t} \right) \Phi(s)\, ds = +\infty \qquad (6.6.25)$$

for any sufficiently large number t_0, where the function $\Phi(t)$ is defined by (6.6.6). Then each solution $u(x, t)$ of problem (6.6.1), (6.6.2) oscillates in the domain G.

A corollary of Theorem 6.6.2 and Theorem 6.6.3 is the following sufficient condition for oscillation of the solutions of problem (6.6.1), (6.6.3).

Theorem 6.6.5

Let conditions (H6.6) hold in addition to the conditions:

$$\lim_{t \to \infty} \inf \int_{t_0}^{t} \left(1 - \frac{s}{t} \right) \Phi_1(s)\, ds = -\infty \qquad (6.6.26)$$

$$\lim_{t \to \infty} \sup \int_{t_0}^{t} \left(1 - \frac{s}{t} \right) \Phi_1(s)\, ds = +\infty \qquad (6.6.27)$$

for any sufficiently large number t_0, where the function $\Phi_1(t)$ is defined by (6.6.14). Then each solution $u(x,t)$ of problem (6.6.1), (6.6.3) oscillates in the domain G.

6.7   Notes and comments to Chapter 6

The results of Section 6.1 are due to Mishev and Bainov [120]. The idea of averaging the solution (6.1.14) comes from Yoshida [156]. The results of Section 6.2 are due to Mishev [115]. Lemma 6.2.3 is a generalization of a result of Zahariev and Bainov [160]. By means of it Theorem 6.2.3 is proved. The main result of this section is contained in Theorem 6.2.4. From Example 6.2.1 it is seen that for neutral equations a new effect arises in comparison with the case when $\lambda(t) \equiv 0$. It is shown that condition (6.2.20) of Theorem 6.2.4 is essential and in general it cannot be replaced by the explicit condition (6.2.21) of Yoshida [156]. The results of Section 6.3 are due to Mishev [116]. The main results are contained in Theorem 6.3.4 and Theorem 6.3.5. The results of Section 6.4 are new and due to Mishev. The main results are formulated in Theorem 6.4.4 and Theorem 6.4.5. The first work devoted to the investigation of the oscillatory properties of the solutions of neutral partial differential equations is due to Mishev and Bainov [122]. Theorem 6.5.4 and Theorem 6.5.5 of Section 6.5 are a generalization of this result and are due to Mishev and Bainov [121], [119]. Some particular cases in which new sufficient conditions are obtained for oscillation of the solutions, are considered in Theorem 6.5.7 and Theorem 6.5.9 by Mishev and Bainov. Lemma 6.5.3 is due to Yoshida [155]. In Theorem 6.5.10 and Theorem 6.5.11 of Yoshida [154] results on the distribution of the zeros of the solutions of neutral nonlinear hyperbolic equations are obtained. Sufficient conditions for oscillation of the solutions are obtained by Yoshida in Corollary 6.5.1 and Corollary 6.5.2. The results of Section 6.6 are due to Mishev [113]. The main results are contained in Theorem 6.6.4 and Theorem 6.6.5.

References

[1] Akhmerov R R, Kamenskii M I, Potapov A S, Rodkina A E and Sadovskii B N 1982 Theory of equations of neutral type *Math. Anal.* (Moscow: VINITI) **19** 55–126 (in Russian)

[2] Aliev A I 1987 *On the Oscillatory Properties of First Order Equations of Neutral Type with Two Deviating Arguments* (VINITI) no 3555-B-87 (in Russian)

[3] Angelov V G and Bainov D D 1987 On the functional differential equations with 'maxima' *Periodica Mathematica Hungarica* **18** (1) 7–15

[4] Arino O and Bourad F On the asymptotic behavior of the solutions of a class of scalar neutral equations generating a monotone semi-flow *J. Diff. Eqs* (to appear)

[5] Arino O and Györi I 1989 Necessary and sufficient condition for oscillation of a neutral differential system with several delays *J. Diff. Eqs* **81** 98–105

[6] Arino O, Györi I and Jawhari A 1984 Oscillation criteria in delay equations *J. Diff. Eqs* **53** 115–23

[7] Bainov D D, Myshkis A D and Zahariev A I 1986 Asymptotic and oscillatory properties of a class of operator-differential inequalities *Annali di Mat. Pura ed Appl.* (IV) **CXLIII** 197–205

[8] Bainov D D, Myshkis A D and Zahariev A I Necessary and sufficient conditions for oscillation of the solutions of linear autonomous functional differential equations of neutral type with distributed delay *J. Math. Anal. Appl.* (to appear)

[9] Bainov D D, Myshkis A D and Zahariev A I 1989 On the oscillatory properties of the solutions of non-linear neutral functional differential equations of second order *Hiroshima Math J.* **19** 203–8

[10] Bainov D D, Myshkis A D and Zahariev A I 1984 Oscillatory properties of the solutions of a class of neutral type integro-differential equations *Bulletin of the Institute of Math. Academia Sinica* **12** 337–42

[11] Bainov D D, Myshkis A D and Zahariev A I 1988 Oscillatory properties of the solutions of linear equations of neutral type *Bull. Austral. Math. Soc.* **38** 255–61

[12] Bainov D D, Myshkis A D and Zahariev A I Oscillatory properties of the solutions of second order linear equations of neutral type with distributed delay (to appear)

[13] Bainov D D, Myshkis A D and Zahariev A I 1987 Sufficient conditions for the existence of bounded non-oscillating solutions of functional differential equations of neutral type *Rendiconti Matematica* **7** 353–9

[14] Bainov D D and Zahariev A I 1984 Oscillating and asymptotic properties of a class of functional differential equations with maxima *Czechoslovak Math. J.* **34** (109) 247–51

[15] Bainov D D, Zahariev A I and Myshkis A D 1984 Oscillatory properties of the solutions of a class of integro-differential equations of neutral type In *IX ICNO* (Kiev: Naukova Dumka) vol 2 39–40 (in Russian)

[16] Bellman R and Cooke K 1963 *Differential-Difference Equations* (New York: Academic Press)

[17] Bykov Ya V and Kultaev T Ch 1983 Oscillation of solutions of a class of parabolic equations *Izv. Acad. Sci. Kirgiz. SSR* **6** 3–9 (in Russian)

[18] Domoshnitskii A I 1986 Constant signs of Cauchy's function and stability of linear equation of neutral type with respect to the right-hand side *Boundary Value Problems* Perm' PPI 44–8 (in Russian)

[19] Domoshnitskii A I Extension of Sturm's theorem to equations with a retarded argument *Differential Equations* **19** 1475–82 (in Russian)

[20] Domoshnitskii A I 1982 On the estimation of the interval of non-oscillation for equations of neutral type *Boundary Value Problems* Perm' PPI 39–42 (in Russian)

[21] Domshlak Yu I 1986 *Comparison Method by Sturm in the Investigation of the Behaviour of Solutions of Differential-Operator Equations* (Elm: Bacou) (in Russian)

[22] Domshlak Yu I 1984 On the distribution of zeros and oscillation of the solutions of first order equations of neutral type *Izv. Acad. Sci. Azerb. SSR, Ser. Phys.-Techn. and Math. Sci.* No 2 24–30 (in Russian)

[23] Domshlak Yu I and Sheikhzamanova L A 1984 On the distribution of zeros of the solutions of integro-differential inequalities of neutral type *Izv. Acad. Sci. Azerb. SSR, Ser. Phys.-Techn. and Math. Sci.* No 3 40–5 (in Russian)

[24] Domshlak Yu I and Sheikhzamanova L A 1985 On the distribution of zeros of the solutions of second order functional differential inequalities of neutral type *Izv. Acad. Sci. Azerb. SSR, Ser. Phys.-Techn. and Math. Sci.* No 4 23–8 (in Russian)

[25] Drakhlin M E 1986 On the oscillatory properties of some functional differential equations *Differential Equations* **22** 396–402 (in Russian)

[26] Driver R D 1984 A mixed neutral system *Nonlinear Anal.: TMA* **8** 155–8

[27] Driver R D 1965 Existence and continuous dependence of solutions on neutral functional-differential equations *Archs. Ration. Mech. Anal.* **19** 149–66

[28] Driver R 1978 *Introduction to Ordinary Differential Equations* (New York: Harper and Row)

[29] El'sgol'ts L E and Norkin S B 1973 *Introduction to the Theory and Application of Differential Equations with Deviating Arguments* (Academic Press)

[30] Erbe L H and Zhang B G 1988 Oscillation for first order linear differential equations with deviating arguments *Differential and Integral Eqs* **1**

[31] Erbe L H and Zhang B G Oscillation of second order neutral differential equations *Bull. Austral. Math. Soc.* (to appear)

[32] Farrell K 1989 Necessary and sufficient conditions for oscillation of neutral equations with real coefficients *J. Math. Anal. Appl.* **140** 251–61

[33] Ferreira J 1986 On the stability of a distributed network *SIAM J. Math. Analysis* **17** 38–45

[34] Ferreira J and Györi I 1987 Oscillatory behavior in linear retarded functional differential equations *J. Math. Anal. Appl.* **128** 332–42

[35] Fite W B 1921 Properties of the solutions of certain functional differential equations *Trans. Amer. Math. Soc.* **22** 311–19

[36] Georgiou D and Kreith K 1985 Functional characteristic initial value problems *J. Math. Anal. Appl.* **107** 414–24

[37] Gopalsamy K 1985 Nonoscillation in a delay-logistic equation *Quart. Appl. Math.* **43** 189–97

[38] Gopalsamy K 1986 Oscillation in a delay-logistic equation *Quart. Appl. Math.* **44** 447–61

[39] Gopalsamy K and Zhang B G 1988 On a neutral delay logistic equation *Dynamics and Stability of Systems* **2** 183–95

[40] Grace S R and Lalli B S 1989 Oscillation and asymptotic behavior of certain second order neutral differential equations *Radovi Mat.* **5** 121–6

[41] Grace S R and Lalli B S 1987 Oscillations of nonlinear second order neutral delay differential equations *Radovi Mat.* **3** 77–84

[42] Grace S R and Lalli B S 1987 Oscillation theorems for certain neutral differential equations *Proceedings of the Eleventh International Conference on Nonlinear Oscillations* Budapest

[43] Graef J R, Grammatikopoulos M K and Spikes P W 1988 Asymptotic properties of solutions of nonlinear neutral delay differential equations of the second order *Radovi Mat.* **4** 133–49

[44] Graef J R, Grammatikopoulos M K and Spikes P W Behavior of the

nonoscillatory solutions of first order neutral delay differential equations (to appear)

[45] Grammatikopoulos M K, Grove E A and Ladas G 1986 Oscillation and asymptotic behavior of neutral differential equations with deviating arguments *Applicable Analysis* **22** 1–19

[46] Grammatikopoulos M K, Grove E A and Ladas G 1987 Oscillation and asymptotic behavior of second order neutral differential equations with deviating arguments *Canadian Math. Soc.* **8** 153–61

[47] Grammatikopoulos M K, Grove E A and Ladas G 1986 Oscillation of first-order neutral delay differential equations *J. Math. Anal. Appl.* **120** 510–20

[48] Grammatikopoulos M K, Ladas G and Meimaridou A 1988 Oscillation and asymptotic behavior of higher order neutral equations with variable coefficients *Chin. Ann. of Math.* **9B** (3) 322–38

[49] Grammatikopoulos M K, Ladas G and Meimaridou A 1985 Oscillations of second order neutral delay differential equations *Radovi Mat.* **1** 267–74

[50] Grammatikopoulos M K, Ladas G and Meimaridou A 1987 Oscillation and asymptotic behavior of second order neutral differential equations *Annali di Matem. Pura ed Appl.* (*IV*), vol CXLVIII 29–40

[51] Grammatikopoulos M K, Ladas G and Sficas Y G 1986 Oscillation and asymptotic behavior of neutral equations with variable coefficients *Radovi Matem.* **2** 279–303

[52] Grammatikopoulos M K, Sficas Y G and Stavroulakis I P 1988 Necessary and sufficient conditions for oscillations of neutral equations with several coefficients *J. Diff. Eqs.* **76** 294–311

[53] Grammatikopoulos M K and Stavroulakis I P Necessary and sufficient conditions for oscillation of neutral equations with deviating arguments. *J. London Math. Soc.* (to appear)

[54] Grammatikopoulos M K and Stavroulakis I P 1989 Oscillations of neutral differential equations University of Ioannina Math. Technical Respect No 165

[55] Grove E A, Kulenovic R S and Ladas G A Myskis-type comparison result for neutral equations (to appear)

[56] Grove E A, Kulenović M R and Ladas G 1987 Sufficient conditions for oscillation and nonoscillation of neutral equations *J. Diff. Eqs.* **68** 373–82

[57] Grove E A and Ladas G 1987 A necessary and sufficient condition for the oscillation of neutral equations *J. Math. Anal. Appl.* **126** 341–54

[58] Grove E A, Ladas G and Schinas J 1988 Sufficient conditions for the oscillation of delay and neutral delay equations *Canad. Math. Bull.* **31** 459–66

[59] Györi I and Witten M Cell population models with a growth rate lag (to appear)

[60] Györi I 1984 Oscillation behaviour of solutions of certain nonlinear and linear delay differential equations *Nonlinear Analysis: TMA* **8** 429–39

[61] Györi I 1986 Oscillation conditions in scalar linear delay differential equations *Bull. Austral. Math. Soc.*, **34** 1–9

[62] Györi I Oscillation of neutral delay differential equations arising in population dynamics *Proceedings of EQUADIFF'87 Conference* Xanthi Greece

[63] Györi I 1989 Oscillations and comparison results in neutral differential equations and their applications to the delay logistic equation *Computers Math. Appl.* **18** 893–906

[64] Györi I 1989 Oscillations of retarded differential equations of the neutral and the mixed type *J. Math. Anal. Appl.* **141** 1–20

[65] Györi I and Ladas G 1988 Oscillations of systems of neutral differential equations *Differential and Integral Eqs* **1** 281–6

[66] Hale J 1987 *Theory of Functional Differential Equations* (New York: Springer)

[67] Headley V B and Swanson C A 1968 Oscillation criteria for elliptic equations *Pacific J. Math.* **27** 501–6

[68] Henry D 1970 Small solutions of linear autonomous functional differential equations *J. Diff. Eqs* **8** (3) 494–501

[69] Hsiang Wu-Teh and Kwong Man Kam 1982 On the oscillation of nonlinear hyperbolic equations *J. Math. Anal. Appl.* **85** 31–45

[70] Hsiang Wu-Teh and Kwong Man Kam 1982 Oscillation of second-order hyperbolic equations with non-integrable coefficients *Proc. Roy. Soc. Edinburgh* **91A** 305–13

[71] Hunt B R and Yorke J A 1984 When all solutions of $x' = -\sum_{i=1}^{n} q_i(t)x(t - T_i(t))$ oscillate *J. Diff. Eqs* **53** 139–45

[72] Ivanov A F 1983 On the oscillation of the solutions of first order differential-difference equations of neutral type Kiev (Preprint *Acad. Sci. Ukr. SSR. Inst. Math* 83.16) (in Russian)

[73] Ivanov A F and Kusano T 1987 On the oscillation of the solutions of a class of differential-functional equations *Ukr. Math. J.* **39** 717–21 (in Russian)

[74] Ivanov A F and Kusano T 1989 Oscillation of the solutions of a class of first order functional differential equations of neutral type *Ukr. Math. J.* **41** 1370–5 (in Russian)

[75] Jaroš J and Kusano T Asymptotic behavior of nonoscillatory solutions of nonlinear functional differential equations of neutral type *Funkcial. Ekvac.* (to appear)

[76] Jaroš J and Kusano T Oscillation properties of first order nonlinear functional differential equations of neutral type (to appear)

[77] Jaroš J and Kusano T Oscillation theory of higher order linear functional differential equations of neutral type *Hiroshima Math. J.* (to appear)

[78] Jaroš J and Kusano T Sufficient conditions for oscillations in higher order linear functional differential equations of neutral type *Japan J. Math.* (to appear)

[79] Kamenskii G A 1958 On the general theory of equations with deviating argument *Dokl. Akad. Nauk SSSR* **120** 697–700

[80] Kiguradze I T 1964 On the oscillation of the solutions of equations $u^{(m)} + a(t)\operatorname{sign} u \cdot |u|^n = 0$ *Math. Studies* **65** 172–87 (in Russian)

[81] Kiguradze I T 1975 Some singular boundary value problems for ordinary differential equations *Izdat. Tbilisk. Univ.* Tbilisi (in Russian)

[82] Kolmanovskii V B and Nosov V R 1986 *Stability of Functional Differential Equations* (Academic Press)

[83] Konstantinov M M and Bainov D D 1972 Theorems for existence and uniqueness of the solution of some differential equations of superneutral type *Publ. Inst. Math. (Beograd)* **14** 75–82

[84] Koplatadze R G and Chanturia T A 1977 On the oscillatory properties of differential equations with a deviating argument *Izdat. Tbilisk. Univ.* Tbilisi (in Russian)

[85] Kreith K 1983 Oscillation properties of weakly time dependent hyperbolic equations *Canad. Math. Bull.* **26**(3) 368–73

[86] Kreith K 1964 Oscillation theorems for elliptic equations *Proc. Amer. Math. Soc.* **15** 341–4

[87] Kreith K 1973 *Oscillation Theory* Lecture Notes in Math. **324** (Springer)

[88] Kreith K 1982 Picone-type theorems for hyperbolic partial differential equations *Pacific J. Math.* **102** 385–95

[89] Kreith K 1983 Picone-type theorems for semidiscrete hyperbolic equations *Proc. Amer. Math. Soc.* **88** 436–8

[90] Kreith K 1969 Sturmian theorems for hyperbolic equations *Proc. Amer. Math. Soc.* **22** 277–81

[91] Kreith K, Kusano T and Yoshida N 1984 Oscillation properties of nonlinear hyperbolic equations *SIAM J. Math. Anal.* **15** 570–8

[92] Kreith K and Ladas G 1985 Allowable delays for positive diffusion processes *Hiroshima Math. J.* **15** 437–43

[93] Kreith K and Pagan G 1983 Qualitative theory for hyperbolic characteristic initial value problems *Proc. Roy. Soc. Edinburgh* **94A** 15–24

[94] Kulenovic M R, Ladas G and Meimaridou A 1987 Necessary and sufficient condition for oscillations of neutral differential equations *J. Austr. Math. Soc. Ser. B* **28** 362–75

[95] Kulenovic M R, Ladas G and Meimaridou A On oscillation of nonlinear delay differential equations *Quart. Appl. Math.* (to appear)

[96] Kusano T and Naito M 1982 Oscillation criteria for a class of perturbed Schrödinger equations *Canad. Math. Bull.* **25** 71–7

[97] Ladas G Linearized oscillations for neutral equations (to appear)

[98] Ladas G 1979 Sharp conditions for oscillations caused by delays *Appl. Anal.* **9** 93–8

[99] Ladas G and Sficas Y G 1988 Asymptotic behaviour of oscillatory solutions *Hiroshima Math. J.* **18** 351–9

[100] Ladas G and Sficas Y G Oscillations of higher order neutral equations 1986 *J. Austral. Math. Soc. Ser. B* **27** 502–11

[101] Ladas G and Sficas Y G 1986 Oscillations of neutral delay differential equations *Canad. Math. Bull.* **29** (4) 438–45

[102] Ladas G, Sficas Y G and Stavroulakis I P 1984 Nonoscillatory functional differential equations *Pacif. J. Math.* **115** 391–8

[103] Ladas G and Stavroulakis I P 1982 On delay differential inequalities of first order *Funkcialaj Ekvacioj* **25** 105–13

[104] Ladas G and Stavroulakis I P 1982 On delay differential inequalities of higher order *Canad. Math. Bull.* **25** 348–54

[105] Ladas G and Stavroulakis I P 1982 Oscillations caused by several retarded and advanced arguments *J. Diff. Eqs* **44** 134–52

[106] Ladas G and Stavroulakis I P 1984 Oscillations of differential equations of the mixed type *J. Math. Phys. Sciences* **18** 245–62

[107] Ladas G and Partheniadis E C 1989 Necessary and sufficient conditions for oscillations of second order neutral equations *J. Math. Anal. Appl.* **138** 214–31

[108] Ladas G, Partheniadis E C and Sficas Y G *Oscillations of second order neutral equations* (to appear)

[109] Ladde G S, Lakshmikantham V and Zhang B G 1987 *Oscillation Theory of Differential Equations with Deviating Arguments* (Marcel Dekker)

[110] Magomedov A R 1977 On some questions of differential equations with maxima *Izv. Acad. Sci. Azerb. SSR, Ser. Phys.-Techn. and Math. Sci.* No 1 104–8 (in Russian)

[111] Magomedov A R 1979 Theorem of existence and uniqueness of the solutions of linear differential equations with maxima *Izv. Acad. Sci. Azerb. SSR, Ser. Phys.-Techn. and Math. Sci.* No 5 116–18 (in Russian)

[112] Mishev D P 1991 Necessary and sufficient conditions for oscillation of neutral type parabolic differential equations *Comptes Rendus Acad. Bulg. Sci.* **44**(3) 11–5

[113] Mishev D P Oscillation of the solutions of hyperbolic differential equations of neutral type with 'maxima' *Godishnik VUZ Appl. Math.* (to appear)

[114] Mishev D P 1991 Oscillation of the solutions of neutral type hyperbolic differential equations *Mathematica Balkanica* **5** 121–8

[115] Mishev D P Oscillation of the solutions of nonlinear parabolic equations of neutral type *Applicable Analysis* (to appear)

[116] Mishev D P Oscillation of the solutions of parabolic differential equations of neutral type with 'maxima' *Godishnik VUZ Appl. Math.* (to appear)

[117] Mishev D P 1986 Oscillatory properties of the solutions of hyperbolic differential equations with maximum *Hiroshima Math. J.* **16** 77–83

[118] Mishev D P and Bainov D D Necessary and sufficient conditions for oscillation of neutral type hyperbolic differential equations (to appear)

[119] Mishev D P and Bainov D D Oscillation of the solutions of nonlinear hyperbolic equations of neutral type *Publicationes Mathematiques* (to appear)

[120] Mishev D P and Bainov D D 1988 Oscillation of the solutions of parabolic differential equations of neutral type *Appl. Math. and Computation* **28** 97–112

[121] Mishev D P and Bainov D D 1986 Oscillation properties of the solutions of a class of hyperbolic equations of neutral type *Funkcialaj Ekvacioj* **29** 213–18

[122] Mishev D P and Bainov D D 1984 Oscillation properties of the solutions of hyperbolic equations of neutral type *Proceedings of the Colloquium on Qualitative Theory of Differential Equations* Szeged 771–80

[123] Myshkis A D 1972 *Linear Differential Equations with Retarded Argument* (Moscow: Nauka) (in Russian)

[124] Myshkis A D 1977 On some problems of the theory of differential equations with a deviating argument *Uspekhi Mat. Nauk* XXXII 2 (194)173–202 (in Russian)

[125] Myshkis A D, Bainov D D and Zahariev A I 1984 Oscillatory and asymptotic properties of a class of operator-differential inequalities *Proc. Roy. Soc. Edinburgh* **96A** 5–13

[126] Nisbet P M and Gurney W S C *Modelling Fluctuating Populations* (New York: Wiley)

[127] Norkin S B 1977 Oscillation of the solutions of differential equations with deviating argument *Differential Equations with Deviating Argument* (Kiev:Naukova Dumka) (in Russian)

[128] Norkin S B 1985 *Second Order Differential Equations with Retarded Argument* (Moscow: Nauka) (in Russian)

[129] Ntouyas S K and Sficas Y G 1983 On the asymptotic behavior of neutral functional differential equations *Arch. Mat.* **41** 352–62

[130] Partheniadis E C 1988 Stability and oscillation of neutral delay differential equations with piecewise constant argument *Differential and Integral Eqs* **1** 459–72

[131] Philos C G 1980 On the existence of nonoscillatory solutions tending to zero at ∞ for differential equations with positive delays *Archiv. Math.* **36** 168–78

[132] Popov E P 1966 *Automatic Regulations and Control* (Moscow: Nauka)

[133] Protter M H 1959 A comparison theorem for elliptic equations *Proc. Amer. Math. Soc.* **10** 296–9

[134] Reid W T 1980 *Sturmian Theory for Ordinary Differential Equations* (New York: Springer)

[135] Ruan J 1986 Oscillations of neutral differential difference equations with several retarded arguments *Sciential Sinica (A)* **10** 1132–44

[136] Ruan J 1988 Oscillatory and asymptotic behavior of first order neutral functional differential equations *J. Central China Normal University* **3**

[137] Ruan J 1989 Oscillatory and asymptotic behavior of n order neutral functional differential equations *Chin. Ann. of Math.* **10B** 143–53

[138] Ruan J 1987 Types and criteria of nonoscillatory solutions for second order linear neutral differential equations *Chinese Ann. Math. Ser.* **A8** 114–24 (in Chinese)

[139] Sficas Y G and Stavroulakis I P 1987 Necessary and sufficient conditions for oscillations of neutral differential equations *J. Math. Anal. Appl.* **123** 494–507

[140] Sheikhzamanova L A 1985 On the distribution of zeros of the solutions of second order functional differential inequalities of neutral type Dep. (VINITI) no 2760-85 (in Russian)

[141] Shevelo V N 1978 *Oscillation of the Solutions of Differential Equations with a Deviating Argument* (Kiev: Naukova Dumka) (in Russian)

[142] Shevelo V N and Varekh N V 1973 On some oscillatory theorems for higher order differential equations *Math. Physics* **13** 183–9 (in Russian)

[143] Shevelo V N, Varekh N V and Gritsai A G 1985 Oscillatory properties of the solutions of systems of differential equations with deviating argument Preprint 85.10 *Kiev Inst. Math. Acad. Sci. Ukr. SSR* (in Russian)

[144] Slemrod M and Infante E F 1972 Asymptotic stability criteria for linear system of difference-differential equations of neutral type and their discrete analogues *J. Math. Anal. Appl.* **38** 399–415

[145] Snow W 1965 Existence, uniqueness and stability for nonlinear differential-difference equations in the neutral case *N.Y.U. Courant Inst. Math. Sci. Rep.* IMM NYU **328**

[146] Stavroulakis I P Oscillation of mixed neutral equations *Hiroshima Math. J.* (to appear)

[147] Sturm C 1836 Sur les équations différentielles linéaires du second ordre *J. Math Pures et Appl.* **1** 106–86

[148] Tramov M I 1975 Conditions for oscillatory solutions of first order differential equations with a delayed argument *Izv. Vyssh. Uchebn. Zaved. Mat.* **19** 92–6 (in Russian)

[149] Tramov M I 1984 On the oscillation of the solutions of partial differential equations with deviating argument *Differential Equations* **20** 721–3 (in Russian)

[150] Vladimirov V S 1981 *Equations of Mathematical Physics* (Moscow: Nauka) (in Russian)

[151] Waltman P 1968 A note on an oscillation criterion for an equation with a functional argument *Canad. Math. Bull.* **11** 593–5

[152] Yoshida N 1979 An oscillation theorem for characteristic initial value problems for nonlinear hyperbolic equations *Proc. Amer. Math. Soc.* **76** 95–100

[153] Yoshida N 1987 Forced oscillations of solutions of parabolic equations *Bull. Austral. Math. Soc.* **36** 289–94

[154] Yoshida N 1990 On the zeros of solutions of hyperbolic equations of neutral type *Differential and Integral Eqs* **3** 155–60

[155] Yoshida N 1987 On the zeros of solutions to nonlinear hyperbolic equations *Proc. Roy. Soc. Edinburgh* **106A** 121–9

[156] Yoshida N 1986 Oscillation of nonlinear parabolic equation with functional arguments *Hiroshima Math. J.* **16** 305–14

[157] Yoshida N 1987 Oscillation properties of solutions of characteristic initial value problems *Proc. International Conference on Nonlinear Oscillations* Budapest 527–30

[158] Zahariev A I and Bainov D D 1988 Integral averaging and oscillation of the solutions of neutral type functional-differential equations *Tamkang J. Math.* **19** 61–7

[159] Zahariev A I and Bainov D D 1986 On some oscillation criteria for a class of neutral type functional differential equations *J. Austral. Math. Soc. Ser. B* **28** 229–39

[160] Zahariev A I and Bainov D D 1980 Oscillating properties of the solutions of a class of neutral type functional differential equations *Bull. Austral. Math. Soc.* **22** 365–72

[161] Zhang B G 1986 A survey of the oscillation of solutions to first order differential equations with deviating arguments *Ann. Diff. Eqs* **2** 65–86

[162] Zhang B G 1989 Oscillation of first order neutral functional differential equations *J. Math. Anal. Appl.* **139** 311–18

[163] Zhanyuan H 1989 On the oscillation of neutral delay differential equations with constant coefficients *Proceedings of the Conference on Differential Equations and Applications* Rousse Bulgaria

Index